T0301370

GUIDELINES FOR
IMPLEMENTING PROCESS SAFETY MANAGEMENT

PUBLICATIONS AVAILABLE FROM THE CENTER FOR CHEMICAL PROCESS SAFETY of the AMERICAN INSTITUTE OF CHEMICAL ENGINEERS

GUIDELINES FOR
IMPLEMENTING PROCESS SAFETY MANAGEMENT
Second Edition

CENTER FOR CHEMICAL PROCESS SAFETY
of the
AMERICAN INSTITUTE OF CHEMICAL ENGINEERS
New York, NY

Published by John Wiley & Sons, Inc., Hoboken, New Jersey.

Published simultaneously in Canada.

Library of Congress Cataloging-in-Publication Data:

Names: American Institute of Chemical Engineers. Center for Chemical Process Safety, author.
Title: Guidelines for implementing process safety management / by: CCPS.
Description: Second edition. | Hoboken, New Jersey : John Wiley & Sons Inc., [2016] | Includes bibliographical references and index.
Identifiers: LCCN 2016024808| ISBN 978-1-118-94948-1 (cloth) | ISBN 978-1-119-24376-2 (epub)
Subjects: LCSH: Chemical processes--Safety measures.
Classification: LCC TP149 .G838 2016 | DDC 660/.2804--dc23 LC record available at https://lccn.loc.gov/2016024808

Printed in the United States of America.

10 9 8 7 6 5 4 3 2 1

This book is one in a series of process safety guidelines and concept books published by the Center for Chemical Process Safety (CCPS). Please go to *www.wiley.com/go/ccps* for a full list of titles in this series.

CONTENTS

FILES ON THE WEB ACCOMPANYING THIS BOOK

Access the following *Guidelines for Implementing Process Safety Management* tools and documents at:

http://www.aiche.org/ccps/implementing-ps-download

Password: Implement2016

- Eli Lilly and Company, PSM Tools and Globally Integrated Process Safety Management (GIPSM) Maps (i.e., PSM system workflows), including:
    1. Process Risk Screening Tool
    2. PSSR Process Description Document
    3. PSSR Map
    4. PSSR Requirements Document
    5. Safety Critical Operations (SCO) Selection Tool
    6. PSI Template
    7. Change Management Response and IAT (Impact Assessment Tool) Comments
    8. MI List – example
    9. MITI (Mechanical Integrity Test and Inspection) Guideline
    10. CPH (Catastrophic Process Hazards) Risk Screening Tool and Worked Example
    11. CPH Program
    12. Serious Injury and Fatality (SIF) Pre-Cursor Guide
    13. Process Hazard Review Process Maps (6)
    14. Process Hazard Review for Capital Projects Maps (5)
- Process Safety Culture Survey (from Baker Panel Report)
- PSM Project Implementation Plan Example
- PSM Software Compilation
- *Recommended Guidelines for Contractor Safety and Health*, Texas Chemical Council Occupational Safety Committee, 2008
- Copies of various global regulations and good industry practices (referenced in Appendix I)

- Integrated Process Safety/HSSE/Business Triangles and Matrices

LIST OF FIGURES

LIST OF TABLES

ACRONYMS AND ABBREVIATIONS

ACC	American Chemistry Council
AIChE	American Institute of Chemical Engineers
ALARP	As Low as Reasonably Practicable
ANSI	American National Standards Institute
API	American Petroleum Institute
ASME	American Society of Mechanical Engineers
ASTM	American Society for Testing and Materials
BMS	Business Management System
CA	Compliance Audits
Capex	Capital Expenditure
CCPS	Center for Chemical Process Safety
CFATS	Chemical Facility Anti-Terrorism Standards
CFR	Code of Federal Regulations
CIAC	Chemical Industry Association of Canada
COMAH	Control of Major Accident Hazards (UK HSE Regulation)
CON	Contractors
COO	Conduct of Operations
CPH	Catastrophic Potential Hazard
CPI	Chemical Process Industry
CSChE	Canadian Society for Chemical Engineering
CWS	Compliance with Standards
EHS	Environmental, Health, and Safety
EP	Employee Participation
EPA	U.S. Environmental Protection Agency
EPR	Emergency Planning and Response
EU	European Union
FMEA	Failure Mode and Effects Analysis
GIPSM	Globally Integrated Process Safety Management
HAZOP	Hazard and Operability
HR	Human Resources
HRO	Higher Risk Operation

HSE	Health and Safety Executive (UK); Health, Safety, and Environmental
HWP	Hot Work Permit
ICI	Imperial Chemical Industries
II	Incident Investigation
IIAR	International Institute of Ammonia Refrigeration
IMS	Information Management System
IP	Intellectual Property
IT	Information Technology
ITPM	Inspection, Testing, and Preventive Maintenance
KPI	Key Performance Indicator
LOPA	Layer of Protection Analysis
MARSEC	Maritime Security
MI	Mechanical Integrity
MITI	Mechanical Integrity Test and Inspection
MKOPSC	Mary Kay O'Connor Process Safety Center
MOC	Management of Change
NFPA	National Fire Protection Association
OD	Operational Discipline
OECD	Organisation for Economic Cooperation and Development
OP	Operating Procedures
OSHA	U.S. Occupational Safety and Health Administration
P&ID	Piping and Instrumentation Diagram
PAR	Performance Assurance Review
PDCA	Plan-Do-Check-Adjust
PHA	Process Hazard Analysis
PIT	Powered Industrial Truck
PSC	Process Safety Competency
PSI	Process Safety Information
PSM	Process Safety Management
PSN	Process Safety Network
PSSR	Pre-startup Safety Review
QRA	Quantitative Risk Assessment
RAGAGEP	Recognized and Generally Accepted Good Engineering Practice
RASCI	Responsible, Accountable, Supports, Consulted, Informed
RBI	Risk Based Inspection
RBPS	Risk Based Process Safety
RCA	Root Cause Analysis
REACH	Registration, Evaluation, Authorisation & Restriction of Chemicals
RMP	Risk Management Program

RP	Recommended Practice
SAWS	State Administration of Work Safety (China)
SCO	Safety Critical Operation
SDS	Safety Data Sheet
SHE	Safety, Health, and Environmental
SHEQ&S	Safety, Health, Environmental, Quality, and Security
SIF	Serious Injury and Fatality
SPI	Safety Performance Indicator
TAPPI	Technical Association of the Pulp and Paper Industry
TQ	Threshold Quantity
TQM	Total Quality Management
TRN	Training
UK	United Kingdom
WHS	Work Health and Safety

GLOSSARY

Acceptable Risk	The average rate of loss that is considered tolerable for a given activity.
Accident	An unplanned event or sequence of events that results in an undesirable consequence.
Accountability	The obligation to explain and answer for one's actions that are related to expectations, objectives, and goals. In this context, those that are accountable for PSM activities are answerable to the one person who has the ultimate responsibility for the program. There may be multiple persons accountable for an activity but only one person with the ultimate responsibility. Accordingly, it is a powerful element of an effective process safety management system.
Action Tracking	A method of logging progress when implementing a task or set of tasks.
Administrative Controls	Procedural mechanisms, such as lockout/tagout procedures, for directing and/or checking human performance on plant tasks.
Adverse Reaction	Undesirable effect of a drug, vaccine, or medical device; it can be as mild as a short-term injection-site irritation or as serious as a life-threatening acute onset of anaphylaxis; also referred to as adverse event.
Alternative Release Scenario (ARS)	The basis for an offsite consequence analysis required by the EPA RMP rule. This release scenario is less conservative, and more likely to occur than the worst-case scenario.

Anecdotal — Verbal evidence that is not supported by other, corroborating evidence. For example, the results of an interview with one person are not the basis for issuing a finding.

Antecedent-behavior-consequence (ABC) Analysis — A human performance analysis tool that examines how human behavior is influenced by previous experiences with similar situations and expectations of reward or punishment.

As Low As Reasonably Practicable (ALARP) — The concept that efforts to reduce risk should be continued until the incremental sacrifice (in terms of cost, time, effort, or other expenditure of resources) is grossly disproportionate to the incremental risk reduction achieved. The term as low as reasonably achievable (ALARA) is often used synonymously.

Asset — Any person, environment, facility, material, information, business reputation, or activity that has positive value to an owner. The asset may have value to an adversary, as well as an owner, although the nature and magnitude of those values may differ.

Asset Integrity — A PSM program element involving work activities that help ensure that equipment is properly designed, installed in accordance with specifications, and remains fit for purpose over its life cycle. Also asset reliability.

Audit — A systematic, independent review to verify conformance with prescribed standards of care using a well-defined review process to ensure consistency and to allow the auditor to reach defensible conclusions.

Audit (Process Safety Audit) — An inspection of a plant or process unit, drawings, procedures, emergency plans, and/or management systems, etc., usually by an independent, impartial team.

Benchmarking — The comparison of current operating practices to internal or external company practices, industry best practices, and regulatory standards.

Catastrophic — A loss with major consequences and unacceptable lasting effects, usually involving significant harm to humans, substantial damage to the environment, and/or loss of community trust with possible loss of franchise to operate.

Catastrophic Release An uncontrolled loss of containment of toxic, reactive, or flammable materials from a process that has the potential for causing onsite or offsite acute health effects, significant environmental effects (e.g., compromise of a public drinking water supply), or significant onsite or offsite property damage.

Causal Factor (CF) Equipment failure or human error that caused an incident or allowed incident consequences to be worse.

Certification Completion of the formal training and qualification requirements specified by applicable codes and standards.

Checklist A list of items requiring verification of completion; typically, a procedure format in which each critical step is marked off (or otherwise acknowledged or verified) as it is performed. Checklists are often appended to procedures that provide a more detailed description of each step, including information regarding hazards, and a more complete description of the controls associated with the hazards. Checklists are also used in conjunction with formal hazard evaluation techniques to ensure thoroughness.

Chemical Any element, chemical compound, or mixture of elements and/or compounds. (OSHA 1994)

Chemical Process Industry The phrase is used loosely to include facilities which manufacture, handle, and use chemicals.

Chemical Reactivity Hazard A situation with the potential for an uncontrolled chemical reaction that can result directly or indirectly in serious harm to people, property, or the environment. The uncontrolled chemical reaction might be accompanied by a temperature increase, pressure increase, gas evolution, or other form of energy release.

Code

Written requirements that affect a facility and/or the process safety requirements that apply to a facility. Codes contain requirements that apply to the design and implementation of management systems, design and operation of process equipment, or similar activities. The difference between a code and a standard is that codes have become part of a law or regulation, and therefore their requirements become mandatory within the jurisdictions that have adopted the code requirements in their laws or regulations. This usually occurs at the state level, but may also occur in local or federal laws or regulations.

Combustible Dust

Any finely divided solid material that is 420 microns or smaller in diameter (material passing through a U.S. No. 40 standard sieve) and presents a fire or explosion hazard when dispersed and <->ignited in air or other gaseous oxidizer.

Combustible Liquid

A term used to classify certain liquids that will burn on the basis of flash points. The National Fire Protection Association (NFPA) defines a combustible liquid as any liquid that has a closed-cup flash point above 100°F (37.8°C) (NFPA 30). There are three subclasses, as follows; Class II liquids have flash points at or above 100°F (37.8°C) but below 140°F (60°C). Class III liquids are subdivided into two additional subclasses; Class IIIA: Those having flash points at or above 140°F (60°C) but below 200°F (93.4°C), Class IIIB: Those having flash points at or above 200°F (93.4°C). The Department of Transportation (DOT) defines degcombustible liquids deg as those having flash points above 140°F (60.5°C) and below 200°F (93.4°C).

Competency

A PSM program element associated with efforts to maintain, improve, and broaden knowledge and expertise.

Computerized Maintenance Management System (CMMS)

Computer software for planning, scheduling, and documenting maintenance activities. A typical CMMS includes work order generation, work instructions, parts and labor expenditure tracking, parts inventories, and equipment histories.

Conduct of Operations (COO) The embodiment of an organization's values and principles in management systems that are developed, implemented, and maintained to (1) structure operational tasks in a manner consistent with the organization's risk tolerance, (2) ensure that every task is performed deliberately and correctly, and (3) minimize variations in performance.

Consequence Analysis The analysis of the expected effects of incident outcome cases, independent of frequency or probability.

Consequences The direct, undesirable result of an accident sequence usually involving a fire, explosion, or release of toxic material. Consequence descriptions may be qualitative or quantitative estimates of the effects of an accident.

Consistency Continued uniformity, during a period or from one period to another.

Continuous Improvement Doing better as a result of regular, consistent efforts rather than episodic or step-wise changes, producing tangible positive improvements either in performance, efficiency, or both. Continuous improvement efforts usually involve a formal evaluation of the status of an activity or management system, along with a comparison to an achievement goal. These evaluation and comparison activities occur much more frequently than formal audits.

Controls Engineered mechanisms and administrative policies and procedures implemented to prevent or mitigate incidents.

Cost Includes tangible items such as money and equipment as well as the operational costs associated with the implementation of risk reduction options. There are also intangible costs such as loss of productivity, moral considerations, political embarrassment, and a variety of others. Costs may be borne by the individuals who are affected or the corporations they work for, or they may involve macroeconomic costs to society.

Cost-benefit Analysis	Part of the management decision-making process in which the costs and benefits of each risk reduction option are compared and the most appropriate alternative is selected.
Covered Process	A process subject to regulatory requirements established under the OSHA PSM standard or the EPA RMP rule.
Critical	Relates to major environment or safety process risks.
Critical Equipment	Equipment, instrumentation, controls, or systems whose malfunction or failure would likely result in a catastrophic release of highly hazardous chemicals, or whose proper operation is required to mitigate the consequences of such release. (Examples are most safety systems, such as area LEL monitors, fire protection systems such as deluge or underground systems, and key operational equipment usually handling high pressures or large volumes.)
Data	A representation of facts, concepts or instructions in a formalized manner suitable for communication, interpretation or processing by human or by "automatic" means. Characters or continuous functions representing information due to know or supposed arrangement.
Determine	To conclude; to reach an opinion consequent to the observation of the fit of sample data within the limit, range, or area associated with substantial conformance, accuracy, or other predetermined standard; to obtain firsthand knowledge of.
Deviation	A process condition outside of established design limits, safe operating limits, or standard operating procedures.
Dow Fire and Explosion Index (F&EI)	A method (developed by Dow Chemical Company) for ranking the relative fire and explosion risk associated with a process. Analysts calculate various hazard and explosion indexes using material characteristics and process data.

Effectiveness The combination of process safety management performance and process safety management efficiency. An effective process safety management program produces the required work products of sufficient quality while consuming the minimum amount of resources.

Element Basic division in a process safety management system that correlates to the type of work that must be done (e.g., management of change [MOC]).

Element Owner The person charged with overall responsibility for overseeing a particular RBPS element. This role is normally assigned to someone who has management or technical oversight of the bulk of the work activities associated with the element, not necessarily someone who performs the work activities on a day-to-day basis.

Emergency Response Plan A written plan which addresses actions to take in case of plant fire, explosion, or accidental chemical release.

Enabling Condition A condition that is not a failure, error, or protection layer but makes it possible for an incident sequence to proceed to a consequence of concern. It consists of a condition or operating phase that does not directly cause the scenario, but that must be present or active in order for the scenario to proceed to a loss event; expressed as a dimensionless probability.

Equipment A piece of hardware which can be defined in terms of mechanical, electrical, or instrumentation components contained within its boundaries.

Equipment Reliability The probability that, when operating under stated environment conditions, process equipment will perform its intended function adequately for a specified exposure period.

Evaluate To reach a conclusion as to significance, worth, effectiveness, or usefulness.

Event An occurrence involving a process that is caused by equipment performance or human action or by an occurrence external to the process.

Facility The physical location where the management system activity is performed.

Failure — An unacceptable difference between expected and observed performance.

Failure Mode and Effects Analysis — A hazard identification technique in which all known failure modes of components or features of a system are considered in turn and undesired outcomes are noted.

Flammable — A gas that can burn with a flame if mixed with a gaseous oxidizer such as air or chlorine and then ignited. The term flammable gas includes vapors from flammable or combustible liquids above their flash points.

Front-line Personnel — The personnel who perform tasks that produce the output of the work group. Front-line personnel include operations and maintenance personnel, engineers, chemists, accountants, shipping clerks, etc.

Hardware — Physical equipment directly involved in performing industrial process measuring and controlling functions, as opposed to computer programs, procedures, rules, and associated documentation.

Hazard — An inherent chemical or physical characteristic that has the potential for causing harm to people, property, or the environment.

Hazard and Operability Study (HAZOP) — A systematic qualitative technique to identify process hazards and potential operating problems using a series of guide words to study process deviations. A HAZOP is used to question every part of a process to discover what deviations from the intention of the design can occur and what their causes and consequences may be. This is done systematically by applying suitable guidewords. This is a systematic detailed review technique, for both batch and continuous plants, which can be applied to new or existing processes to identify hazards.

Hazard Evaluation Identification of individual hazards of a system, determination of the mechanisms by which they could give rise to undesired events, and evaluation of the consequences of these events on health (including public health), environment and property. Uses qualitative techniques to pinpoint weaknesses in the design and operation of facilities that could lead to incidents.

Hazard Identification and Risk Analysis (HIRA) A collective term that encompasses all activities involved in identifying hazards and evaluating risk at facilities, throughout their life cycle, to make certain that risks to employees, the public, or the environment are consistently controlled within the organization's risk tolerance.

Hazardous Chemical A material that is toxic, reactive, or flammable and is capable of causing a process safety incident if released. Also hazardous material.

Hazardous Material In a broad sense, any substance or mixture of substances having properties capable of producing adverse effects to the health or safety of human beings or the environment. Material presenting dangers beyond the fire problems relating to flash point and boiling point. These dangers may arise from, but are not limited to, toxicity, reactivity, instability, or corrosivity

Highly Hazardous Chemical A material that is toxic, reactive, or flammable and is capable of causing a process safety incident if released. These materials are included in OSHA's PSM standard, 29 CFR 1901.119.

Hot Work Any operation that uses flames or can produce sparks (e.g., welding).

Human Factors A discipline concerned with designing machines, operations, and work environments so that they match human capabilities, limitations, and needs. Includes any technical work (engineering, procedure writing, worker training, worker selection, etc.) related to the human factor in operator-machine systems.

Impact A measure of the ultimate loss and harm of a loss event. Impact may be expressed in terms of numbers of injuries and/or fatalities; extent of environmental damage; and/or magnitude of losses such as property damage, material loss, lost production, market share loss, and recovery costs.

Implementation Completion of an action plan associated with the outcome of the process of resolving audit findings, incident investigation team recommendations, risk analysis team recommendations, and so forth. Also, the establishment or execution of PSM program element work activities.

Incident An event, or series of events, resulting in one or more undesirable consequences, such as harm to people, damage to the environment, or asset/business losses. Such events include fires, explosions, releases of toxic or otherwise harmful substances, and so forth.

Incident Investigation A systematic approach for determining the causes of an incident and developing recommendations that address the causes to help prevent or mitigate future incidents.

Incompatible The term can refer to any undesired results occurring when substances are combined. In the context of this publication, it refers to incompatible substances giving an undesired chemical reaction when combined, posing a chemical reactivity hazard under a defined scenario.

Infrastructure The basic facilities, services, and installations needed for the functioning of a site such as transportation and communications systems, water and power lines, and public institutions, including emergency response organizations.

Inherent Safety A concept or an approach to safety that focuses on eliminating or reducing the hazards associated with a set of conditions.

Initiating Event
The minimum combination of failures or errors necessary to start the propagation of an incident sequence. It can be comprised of a single initiating cause, multiple causes, or initiating causes in the presence of enabling conditions. The term initiating event is the usual term employed in layer of protection analysis to denote an initiating cause or, where appropriate, an aggregation of initiating causes with the same immediate effect, such as "BPCS failure resulting in high reactant flow."

Injury
Physical harm or damage to a person resulting from traumatic contact between the body and an outside agency or exposure to environmental factors.

Inspection, Testing, and Preventive Maintenance (ITPM)
Scheduled proactive maintenance activities intended to (1) assess the current condition and/or rate of degradation of equipment, (2) test the operation/functionality of equipment, and/or (3) prevent equipment failure by restoring equipment condition.

Interview
Questioning, both formally and informally, facility personnel or other individuals in order to obtain an understanding of the plant's operations and performance.

Job Safety Analysis (JSA)
A procedure that systematically identifies (1) job steps, (2) specific hazards associated with each job step, and (3) safe job procedures associated with each step to minimize accident potential. Also called job hazard analysis

Knowledge (or Process Safety Knowledge)
Knowledge is related to information, which is often associated with policies, and other rule-based facts. It includes work activities to gather, organize, maintain, and provide information to other process safety elements. Process safety knowledge primarily consists of written documents such as hazard information, process technology information, and equipment-specific information.

Lagging Indicators
Outcome-oriented metrics, such as incident rates, downtime, quality defects, or other measures of past performance.

Lagging Metric
A retrospective set of metrics based on incidents that meet an established threshold of severity.

Layer of Protection Analysis (LOPA)
An approach that analyzes one incident scenario (cause-consequence pair) at a time, using predefined values for the initiating event frequency, independent protection layer failure probabilities, and consequence severity, in order to compare a scenario risk estimate to risk criteria for determining where additional risk reduction or more detailed analysis is needed. Scenarios are identified elsewhere, typically using a scenario-based hazard evaluation procedure such as a HAZOP study.

Leading Indicators
Process-oriented metrics, such as the degree of implementation or conformance to policies and procedures, that support the PSM program management system and has the capability of predicting performance.

Leading Metric
A forward-looking set of metrics that indicate the performance of the key work processes, operating discipline, or layers of protection that prevent incidents.

Lessons Learned
Applying knowledge gained from past incidents in current practices.

Life Cycle
The stages that a physical process or a management system goes through as it proceeds from birth to death. These stages include conception, design, deployment, acquisition, operation, maintenance, decommissioning, and disposal.

Lockout/Tagout
A safe work practice in which energy sources are positively blocked away from a segment of a process with a locking mechanism and visibly tagged as such to help ensure worker safety during maintenance and some operations tasks.

Loss Event

Point in time in an abnormal situation when an irreversible physical event occurs that has the potential for loss and harm impacts. Examples include release of a hazardous material, ignition of flammable vapors or ignitable dust cloud, and overpressurization rupture of a tank or vessel. An incident might involve more than one loss event, such as a flammable liquid spill (first loss event) followed by ignition of a flash fire and pool fire (second loss event) that heats up an adjacent vessel and its contents to the point of rupture (third loss event). Generally synonymous with hazardous event.

Management of Change

A management system to identify, review, and approve all modifications to equipment, procedures, raw materials, and processing conditions, other than replacement in kind, prior to implementation to help ensure that changes to processes are properly analyzed (for example, for potential adverse impacts), documented, and communicated to employees affected.

Management Review

A PSM program element that provides for the routine evaluation of other PSM program management systems/elements with the objective of determining if the element under review is performing as intended and producing the desired results as efficiently as possible. It is an ongoing due diligence review by management that fills the gap between day-to-day work activities and periodic formal audits.

Management System

A formally established set of activities designed to produce specific results in a consistent manner on a sustainable basis.

Mechanical Integrity

A management system focused on ensuring that equipment is designed, installed, and maintained to perform the desired function.

Mechanical Integrity Program

A program to ensure that process equipment and systems are and remain mechanically suitable for operation. It involves inspection, testing, upgrading and repairs of equipment, as well as written procedures to maintain ongoing integrity of equipment.

Methodology The use of a combination of two or more incident investigation tools to analyze the evidence and determine the root causes of the incident.

Metrics Leading and lagging measures of process safety management efficiency or performance. Metrics include predictive indicators, such as the number of improperly performed line-breaking activities during the reporting period, and outcome-oriented indicators, such as the number of incidents during the reporting period.

Mitigate Reduce the impact of a loss event.

Mitigation Lessening the risk of an accident event sequence by acting on the source in a preventive way by reducing the likelihood of occurrence of the event, or in a protective way by reducing the magnitude of the event and/or the exposure of local persons or property.

Near Miss An event in which an accident (that is, property damage, environmental impact, or human loss) or an operational interruption could have plausibly resulted if circumstances had been slightly different.

Normal Operations Any process operations intended to be performed between startup and shutdown to support continued operation within safe upper and lower operating limits.

Normalization of Deviance A gradual erosion of standards of performance as a result of increased tolerance of nonconformance. Also normalization of deviation.

Observation The noting and recording of information to support findings. Also field observation.

Operating Instructions A series of sequential written details describing how to operate equipment.

Operating Limits The values or ranges of values within which the process parameters normally should be maintained when operating. These values are usually associated with preserving product quality or operating the process efficiently; however, they may also incorporate the safe upper and lower limits of the process, or other important limits.

Operating Procedures Written, step-by-step instructions and information necessary to operate equipment, compiled in one document, including operating instructions, process descriptions, operating limits, chemical hazards, and safety equipment requirements.

Operational Discipline (OD) The performance of all tasks correctly every time. Good OD results in performing the task the right way every time. Individuals demonstrate their commitment to process safety through OD. OD refers to the day-to-day activities carried out by all personnel. OD is the execution of the COO system by individuals within the organization.

Operator An individual responsible for monitoring, controlling, and performing tasks as necessary to accomplish the productive activities of a system. Operator is also used in a generic sense to include people who perform a wide range of tasks (e.g., readings, calibration, incidental maintenance, manage loading/unloading, and storage of hazardous materials).

Organizational Culture The common set of values, behaviors, and norms at all levels in a facility or in the wider organization that affect the operation of the facility.

OSHA Process Safety Management (OSHA PSM) A U.S. regulatory standard that requires use of a 14-element management system to help prevent or mitigate the effects of catastrophic releases of chemicals or energy from processes covered by the regulations (49 CFR 1910.119).

Parameter A quantity describing the relation of variables within a given system. Note: A parameter may be constant or depend on the time or the magnitude of some system variables.

Peer Review A series of informal reviews by, and at the discretion of, individual members of the matrix team, as well as more formal reviews (P&ID hazard reviews) held by the entire project matrix or hazard review team in accordance with corporate standards.

Performance A measure of the quality or utility of PSM program work products and work activities.

Performance Assurance A formal management system that requires workers to demonstrate that they understand a training module and can apply the training in practical situations. Performance assurance is normally an ongoing process to (1) ensure that workers meet performance standards and maintain proficiency throughout their tenure in a position and (2) help identify tasks for which additional training is required.

Performance Indicators See metrics.

Performance Measure A metric used to monitor or evaluate the operation of a program activity or management system.

Piping and Instrumentation Diagram (P&ID) A diagram that shows the details about the piping, vessels, and instrumentation.

Plan-Do-Check-Adjust (PDCA) Approach A four-step process for quality improvement. In the first step (Plan), a way to bring about improvement is developed. In the second step (Do), the plan is carried out. In the third step (Check), what was predicted is compared to what was observed in the previous step. In the last step (Adjust), plans are revised to eliminate performance gaps. The PDCA cycle is sometimes referred to as (1) the Shewhart cycle because Walter A. Shewhart discussed the concept in his book entitled *Statistical Method from the Viewpoint of Quality Control* or 2) the Deming cycle because W. Edwards Deming introduced the concept in Japan; the Japanese subsequently called it the Deming cycle. It is also called the Plan-Do-Study-Act (PDSA) cycle.

Pre-Startup Safety Review (PSSR) A systematic and thorough check of a process prior to the introduction of a highly hazardous chemical to a process. The PSSR must confirm the following: construction and equipment are in accordance with design specifications; safety, operating, maintenance, and emergency procedures are in place and are adequate; a process hazard analysis has been performed for new facilities and recommendations have been resolved or implemented before startup; modified facilities meet the management of change requirements; and training of each employee involved in operating a process has been completed.

Prevention The process of eliminating or preventing the hazards or risks associated with a particular activity. Prevention is sometimes used to describe actions taken in advance to reduce the likelihood of an undesired event.

Preventive Maintenance Maintenance that seeks to reduce the frequency and severity of unplanned shutdowns by establishing a fixed schedule of routine inspection and repairs.

Preventive Measures Measures taken at the initial stages of a runaway to avoid further development of the runaway or to reduce its final effects.

Probability The expression for the likelihood of occurrence of an event or an event sequence during an interval of time, or the likelihood of success or failure of an event on test or on demand. Probability is expressed as a dimensionless number ranging from 0 to 1.

Procedures Written step-by-step instructions and associated information (cautions, notes, warnings) that describe how to safely perform a task.

Process A broad term that includes the equipment and technology needed for petrochemical production, including reactors, tanks, piping, boilers, cooling towers, refrigeration systems, etc.

Process Area An area containing equipment (e.g. pipes, pumps, valves, vessels, reactors, and supporting structures) intended to process or store materials with the potential for explosion, fire, or toxic material release.

Process Flow Diagram A diagram that shows the material flow from one piece of equipment to the other in a process. It usually provides information about the pressure, temperature, composition, and flow rate of the various streams, heat duties of exchangers, and other such information pertaining to understanding and conceptualizing the process.

Process Hazard Analysis	An organized effort to identify and evaluate hazards associated with processes and operations to enable their control. This review normally involves the use of qualitative techniques to identify and assess the significance of hazards. Conclusions and appropriate recommendations are developed. Occasionally, quantitative methods are used to help prioritized risk reduction.
Process Safety	A disciplined framework for managing the integrity of operating systems and processes handling hazardous substances by applying good design principles, engineering, and operating practices. It deals with the prevention and control of incidents that have the potential to release hazardous materials or energy. Such incidents can cause toxic effects, fire, or explosion and could ultimately result in serious injuries, property damage, lost production, and environmental impact.
Process Safety Competency	The combination of knowledge, skill, expertise, and training needed to deem someone as well-qualified and capable relating to process safety.
Process Safety Culture	The common set of values, behaviors, and norms at all levels in a facility or in the wider organization that affect process safety.
Process Safety Incident/Event	An event that is potentially catastrophic, i.e., an event involving the release/loss of containment of hazardous materials that can result in large-scale health and environmental consequences.
Process Safety Information (PSI)	Physical, chemical, and toxicological information related to the chemicals, process, and equipment. It is used to document the configuration of a process, its characteristics, its limitations, and as data for process hazard analyses.
Process Safety Management (PSM)	A management system that is focused on prevention of, preparedness for, mitigation of, response to, and restoration from catastrophic releases of chemicals or energy from a process associated with a facility.
Process Safety Metric	A standard of measurement or indicator of process safety management efficiency or performance.

Process Safety Management Systems — Comprehensive sets of policies, procedures, and practices designed to ensure that barriers to episodic incidents are in place, in use, and effective.

Protocol (Audit) — A document that organizes audit procedures into a general sequence of audit steps and describes the actions to be taken by the auditor.

PSM Audit — An activity to determine and status and quality of a PSM program. This term is not used to describe an audit performed exclusively in response to OSHA's PSM standard, but to an audit of any PSM program.

Qualitative — Based primarily on description and comparison using historical experience and engineering judgment, with little quantification of the hazards, consequences, likelihood, or level of risk.

Quality Assurance — A planned, systematic pattern of actions necessary to provide suitable confidence that a system or component will perform satisfactory in actual operation. A systematic pattern of actions throughout design and production, to ensure confidence in a product's conformance with specifications. A set of systematic actions intended to provide confidence that a product or service will continually fulfill a defined need.

Quantitative Risk Assessment (QRA) — The use of quantitative risk analysis results to make decisions, either through relative ranking of risk reduction strategies or through comparison with risk targets.

Reactive Chemical — A substance that can pose a chemical reactivity hazard by readily oxidizing in air without an ignition source (spontaneously combustible or peroxide forming), initiating or promoting combustion in other materials (oxidizer), reacting with water, or self-reacting (polymerizing, decomposing or rearranging). Initiation of the reaction can be spontaneous, by energy input such as thermal or mechanical energy, or by catalytic action increasing the reaction rate.

Reactivity — The relative tendency of a substance to undergo chemical reaction (low, medium, or high).

Recognized and Generally Accepted Good Engineering Practice (RAGAGEP)
Recognized and generally accepted good engineering practices (RAGAGEPs) are the basis for engineering, operation, or maintenance activities and are themselves based on established codes, standards, published technical reports or recommended practices (RP), or similar documents. RAGAGEP details generally approved ways to perform specific engineering, inspection or mechanical integrity activities, such as fabricating a vessel, inspecting a storage tank, or servicing a relief valve.

Recommendation
A suggested course of action intended to prevent the occurrence (or recurrence) of an accident event sequence, or to mitigate its consequences.

Reliability
The probability that an item is able to perform a required function under stated conditions for a stated period of time or for a stated demand.

Replacement-in-kind (RIK)
An item (equipment, chemical, procedure, etc.) that meets the design specification of the item it is replacing. This can be an identical replacement or any other alternative specifically provided for in the design specification, as long as the alternative does not in any way adversely affect the use of the item or associated items.

Resolution
Management's determination of what needs to be done in response to an audit finding (and/or associated recommendation), incident investigation team recommendation, risk analysis team recommendation, and so forth. During the resolution step, management accepts, rejects for cause, or modifies each recommendation. If the recommendation is accepted, an action plan for its implementation will typically be identified as part of the resolution. (See implementation.)

Resources
The labor effort, capital and operating costs, and other inputs that must be provided to execute work activities and produce work products.

Response
A security strategy to neutralize the adversary or to evacuate, shelter in place, call local authorities, control a release, or take other mitigation actions.

Responsibility The single person who has been assigned and has accepted the ultimate accountability for the development and or implementation a program, its separate activities, as well as its success or failure. There can be only one person with the ultimate responsibility for something. Although accountability enters into this definition, that term is used separately in this book.

Responsible Care® An initiative implemented by the Chemical Manufacturers Association (CMA) in 1988 to assist in leading chemical processing industry companies in ethical ways that increasingly benefit society, the economy, and the environment while adhering to 10 key principles.

Risk A measure of human injury, environmental damage, or economic loss in terms of both the incident likelihood and the magnitude of the loss or injury. A simplified version of this relationship expresses risk as the product of the likelihood and the consequences (i.e., Risk = Consequence x Likelihood) of an incident.

Risk Analysis The estimation of scenario, process, facility and/or organizational risk by identifying potential incident scenarios, then evaluating and combining the expected frequency and impact of each scenario having a consequence of concern, then summing the scenario risks if necessary to obtain the total risk estimate for the level at which the risk analysis is being performed.

Risk Assessment The process by which the results of a risk analysis (i.e., risk estimates) are used to make decisions, either through relative ranking of risk reduction strategies or through comparison with risk targets.

Risk Based Approach A quantitative risk assessment methodology used for building siting evaluation that takes into consideration numerical values for both the consequences and frequencies of explosion, fire, or toxic material release.

Risk Based Inspection (RBI)	A risk assessment and management process that is focused on loss of containment of pressurized equipment in processing facilities, due to material deterioration. These risks are managed primarily through equipment inspection.
Risk Based Process Safety (RBPS)	The Center for Chemical Process Safety's process safety management system approach that uses risk based strategies and implementation tactics that are commensurate with the risk based need for process safety activities, availability of resources, and existing process safety culture to design, correct, and improve process safety management activities.
Risk Management	The systematic application of management policies, procedures, and practices to the tasks of analyzing, assessing, and controlling risk in order to protect employees, the general public, the environment, and company assets, while avoiding business interruptions. Includes decisions to use suitable engineering and administrative controls for reducing risk.
Risk Management Program (RMP) Rule	EPA's accidental release prevention rule, which requires covered facilities to prepare, submit, and implement a risk management plan.
Risk Matrix	A tabular approach for presenting risk tolerance criteria, typically involving graduated scales of incident likelihood on the Y-axis and incident consequences on the X-axis. Each cell in the table (at intersecting values of incident likelihood and incident consequences) represents a particular level of risk.
Risk Reduction	Development, comparison, and selection of options to reduce risk to a target level, if needed, or as needed.
Risk Tolerance	The maximum level of risk of a particular technical process or activity that an individual or organization accepts to acquire the benefits of the process or activity.

Root Cause Analysis (RCA) A formal investigation method that attempts to identify and address the management system failures that led to an incident. These root causes often are the causes, or potential causes, of other seemingly unrelated incidents. Identifies the underlying reasons the event was allowed to occur so that workable corrective actions can be implemented to help prevent recurrence of the event (or occurrence of similar events).

Root Causes Management system failures, such as faulty design or inadequate training, that led to an unsafe act or condition resulting in an incident; underlying cause. If the root causes were removed, the particular incident would not have occurred.

Safe Operating Limits Limits established for critical process parameters, such as temperature, pressure, level, flow, or concentration, based on a combination of equipment design limits and the dynamics of the process.

Safe Upper and Lower Limits The safe upper and lower limits refer to equipment design limits, not quality-related operating limits. Sometimes these values are referred to as design limits (e.g., design pressure, design temperature).

Safe Work Practices An integrated set of policies, procedures, permits, and other systems that are designed to manage risks associated with nonroutine activities such as performing hot work, opening process vessels or lines, or entering a confined space.

Safety The expectation that a system does not, under defined conditions, lead to a state in which human life, economics or environment are endangered.

Safety Instrumented System (SIS) The instrumentation, controls, and interlocks provided for safe operation of the process.

Safety System Equipment and/or procedures designed to limit or terminate an incident sequence, thus avoiding a loss event or mitigating its consequences.

Sampling Selecting a portion of a large population of data or information to determine the accuracy, representativeness, or characteristics of the entire population.

Scenario A detailed description of an unplanned event or incident sequence that results in a loss event and its associated impacts, including the success or failure of safeguards involved in the incident sequence.

Screening Tool A simplified dispersion model with limited capabilities, suitable for screening-level studies.

Serious Injury The classification for an occupational injury which includes (a) all disabling work injuries and (b) nondisabling work injuries as follows: (1) eye injuries requiring treatment by a physician, (2) fractures, (3) injuries requiring hospitalization, (4) loss of consciousness, (5) injuries requiring treatment by a doctor, and (6) injuries requiring restriction of motion or work, or assignment to another job.

Severity The maximum credible consequences or effects, assuming no safeguards are in place.

Should In this book the word "should" has been used to refer to action or guidance that is not mandatory. This has been applied to both the compliance and related audit criteria. The reason the compliance criteria are prefaced by should rather than shall, must, or other imperative terms is because the regulations described in this book that govern PSM programs from which the compliance criteria derived are performance-based in nature. Consequently, there may be multiple pathways to successful compliance and it is not the intent of this book to specify one method of compliance as being preferred or better than another, even inadvertently.

Shutdown (S/D) A process by which an operating plant or system is brought to a safe and nonoperating mode.

Siting The process of locating a complex, site, plant, or unit.

Software (S/W) Programs, procedures, rules, and associated documentation required for the operating and/or maintenance of a digital system. Computer programs, routines, programming languages, and systems. The collection of related utility, assembly, and other programs that are desirable for properly presenting a given machine to a user, including detailed procedures to be followed, whether expressed as programs for a computer or as procedures for an operator or other person; documents, including hardware manuals and drawings, computer program listing, and diagrams, etc.; and items such as those listed above, as contrasted with hardware.

Stakeholder Individuals or organizations that can (or believe they can) be affected by the facility's operations, or who are involved with assisting or monitoring facility operations.

Standards The PSM program element, Compliance with Standards, that helps identify, develop, acquire, evaluate, disseminate, and provide access to applicable standards, codes, regulations, and laws that affect a facility and/or the process safety requirements applicable to a facility. More generally, standards also refers to requirements promulgated by regulators, professional or industry-sponsored organizations, companies, or other groups that apply to the design and implementation of management systems, design, and operation of process equipment, or similar activities.

Sustainability Meeting the needs of the present without compromising the ability of future generations to meet their own needs.

System A collection of people, machines, and methods organized to accomplish a set of specific functions.

Testing Verifying that the sampled information is valid. Testing can be performed by retracing data or information (i.e., physically checking against the status of the sampled information against equipment, operations, etc.), independent computation of results, and confirmation using another source of data or information.

Timely — Unless a different definition or explanation of this term is provided in a chapter within a specific context, timely shall mean the following: the resolution or implementation of recommendations, action items, and other follow-up activities are promptly determined, performed, or conducted. This means that they are completed in a reasonable time period given the complexity of the actions or activities decided upon and their difficulty of implementation, and that the timing should be evaluated on a case-by-case basis.

Tolerance — A measure of the uncertainty arising from the physical and the environmental differences between members of differing equipment populations when failure rate data is aggregated to produce a final generic data set.

Toxic Hazard — In the context of these guidelines, a measure of the danger posed to living organisms by a toxic agent, determined not only by the toxicity of the agent itself, but also by the means by which it may be introduced into the subject organisms under prevailing conditions.

Training — Practical instruction in job and task requirements and methods. Training may be provided in a classroom or at the workplace, and its objective is to enable workers to meet some minimum initial performance standards, to maintain their proficiency, or to qualify them for promotion to a more demanding position.

Transparency — Openness of an organization with regard to sharing information about how it operates.

Turnaround — A scheduled shutdown period when planned inspection, testing, and preventive maintenance, as well as corrective maintenance such as modifications, replacements, or repairs, is performed.

Uncertainty — A measure, often quantitative, of the degree of doubt or lack of certainty associated with an estimate of the true value of a parameter.

Underlying Causes — Actual root causes.

Variable A quantity or condition whose value is subject to change and can usually be measured. A language object that may take different values, one at a time. Note: The values of a variable are usually restricted to a certain data type.

Variation A change in data, process parameter, or human behavior. Within prescribed limits, changes in data, process parameters, and human behavior are anticipated and acceptable. Variation outside established limits is called deviation.

Verification A wide variety of activities that can be employed to increase confidence in the audit data, including evaluating the application of, and adherence to, laws, regulations, policies and procedures, standards, and management directives; certifying the validity of data and reports; and evaluating the effectiveness of management systems.

Verify To confirm the truth, accuracy, or correctness of, by competent examination; to substantiate.

Vulnerability Any weakness that can be exploited by an adversary to gain access to an asset.

Worst-case Scenario (WCS) The basis for an offsite consequence analysis required by the EPA RMP rule. This intentionally conservative accident scenario assumes the release of the entire inventory of a vessel, under the most unfavorable conditions, and with the failure of most protective features.

Written Program A description of a management system that defines important aspects such as purpose and scope, roles and responsibilities, tasks and procedures, necessary input information, anticipated results and work products, personnel qualifications and training, activity triggers, desired schedule and deadlines, necessary resources and tools, continuous improvement, management review, and auditing.

ACKNOWLEDGMENTS

The American Institute of Chemical Engineers (AIChE) and the Center for Chemical Process Safety (CCPS) express their appreciation and gratitude to all members of the Implementing Process Safety Management Subcommittee and their CCPS member companies for their generous support and technical contributions in the preparation of these *Guidelines*. The AIChE and CCPS also express their gratitude to the team of authors from ABSG Consulting Inc.

IMPLEMENTING PROCESS SAFETY MANAGEMENT SUBCOMMITTEE MEMBERS:

Robert Stankovich	Committee Chair, Eli Lilly and Company
Abdulrehman Aldeeb	Siemens Energy, Inc.
Steve Arendt	ABSG Consulting Inc.
Rajesh Bhatkhande	Toyo Engineering India, Ltd
Jonas Duarte	Chemtura Corporation
William Fink	Sage Environmental Consulting
David Guss	Nexen, Inc.
Dennis Hickman	Flint Hills Resources (retired)
Shahryar Khajehnajafi	SAFER Systems
Neil Maxson	Bayer Material Science
Georges Melhem	ioMosaic
Cathy Pincus	ExxonMobil
Sara Saxena	BP
Adrian Sepeda	CCPS Fellow

CCPS Staff Consultant: Dan Sliva

CCPS wishes to acknowledge the many contributions of the ABSG Consulting Inc. staff members who wrote this book, especially the principal author James R. Thompson and his colleagues Steve Arendt, Rick Curtis, Marney Gillmore, and Steve Goewert. Document editing assistance from Leslie Adair is gratefully acknowledged as well.

Before publication, all CCPS books are subjected to a thorough peer review process. CCPS gratefully acknowledges the thoughtful comments and suggestions of the peer reviewers. Their work enhanced the accuracy and clarity of these guidelines.

Peer Reviewers:

Mervyn Carneiro	Eli Lilly and Company
Joe Chandler	Flint Hills Resources
Valery Hill	Eli Lilly and Company
Francisco Justiniano	Eli Lilly and Company
Claire Linford	aeSolutions
Donny Maclean	SBM Offshore
Mark Manton	ABSG Consulting Inc.
William Marshall	Eli Lilly and Company
Louis Martin	Bayer Material Science
Jost Andreas Menne	Chemtura Corporation
Darrin Miletello	LyondellBasell
Pamela Nelson	Cytec Industries
Brian Riesing	Eli Lilly and Company
Robert Stankovich	Eli Lilly and Company

PREFACE

The American Institute of Chemical Engineers (AIChE) has been closely involved with process safety and loss control issues in the chemical and allied industries for more than four decades. Through its strong ties with process designers, constructors, operators, safety professionals, and members of academia, AIChE has enhanced communications and fostered continuous improvement of the industry's high safety standards. AIChE publications and symposia have become information resources for those devoted to process safety and environmental protection.

AIChE created the Center for Chemical Process Safety (CCPS) in 1985 after the chemical disasters in Mexico City, Mexico, and Bhopal, India. The CCPS is chartered to develop and disseminate technical information for use in the prevention of major chemical accidents. The center is supported by more than 150 chemical process industry (CPI) sponsors who provide the necessary funding and professional guidance to its technical committees. The major product of CCPS activities has been a series of guidelines to assist those implementing various elements of a process safety and risk management system. This book is part of that series.

The CCPS Technical Steering Subcommittee overseeing this guideline was chartered to review and update the 1994 CCPS book, *Guidelines for Implementing Process Safety Management Systems*. This guideline has been written to reflect the lessons learned about implementing process safety management (PSM) since that original publication, and provide guidance and a road map for the possible PSM implementation situations of:

- implementing a new PSM system,
- implementing new elements into an existing system, or
- improving an existing PSM element or system,

at least one of which may apply to companies new to PSM or to those with mature PSM systems. In addition, this guideline provides practical examples (see the appendices) and tools (see the files on the Web accompanying the book) to aid in PSM implementation.

In this guideline, the committee uses and references the "next generation" PSM system framework published in the 2007 CCPS book, *Guidelines for Risk*

Based Process Safety. The risk based process safety (RBPS) approach recognizes that all hazards and risks are not equal; consequently, it advocates that more resources should be focused on more significant hazards and higher risks. The approach is built on four pillars:

1. Commit to process safety
2. Understand hazards and risk
3. Manage risk
4. Learn from experience

These pillars are further divided into 20 RBPS elements. The 20 RBPS elements built and expanded upon the original 12 elements proposed in *Guidelines for Technical Management of Chemical Process Safety* (1989) and further refined in *Plant Guidelines for Technical Management of Chemical Process Safety* (1992). Thus, they reflect 15 years of PSM implementation experience and well-established best practices from a variety of industries.

RBPS also stresses the principle that PSM management systems should be simplified to the lowest order of complexity while maintaining a fitness-for-purpose objective. Consequently, issues to consider when determining the degree of management system rigor required include:

- the perception of the complexity, hazard, and risk involved with the process, facility, and/or organization;
- the demand for the management system results and the resources required to deliver them; and
- the current facility and/or company process safety culture.

With fitness-for-purpose in mind, the PSM management systems can then be designed, implemented, and maintained to correct and/or improve the system activities.

Therefore, this guideline continues CCPS's efforts to encourage the adoption of a risk based approach to managing process safety in the chemical and allied process industries, so that it becomes an integral part of the effort to continually improve the already impressive process safety performance of these industries.

Finally, this guideline also addresses the important related topics of:

- determining process safety implementation and performance status,
- preparing for PSM system change,
- integrating PSM/HSE with the business management system, and
- managing process safety performance.

1

INTRODUCTION

Companies have been implementing process safety management (PSM) systems for over 25 years. A variety of PSM structures have been used – some based upon regulatory requirements and many more based upon evolving industry good practices. These PSM systems are designed to manage the hazards and risks associated with processes using hazardous chemicals or energy. Management of these aspects requires a PSM system to focus on nurturing the performance of equipment and people throughout the life cycle of their deployment in a facility. The adoption of PSM systems has gone global, offering many new opportunities to improve upon implementation practices of the past.

Moreover, in spite of best efforts and many opportunities for learning lessons, companies are challenged with continually improving process safety performance and efficiency, along with managing all of the other important aspects that a company must concern itself with to be safe and profitable (e.g., occupational safety, environmental, security, economic competitiveness, sustainability). Some companies face the challenge of initial implementation or continual improvement by recognizing that ultimately it is people who must perform – executives, management, staff, operations, maintenance, and contractors – whether it is in designing or executing the intended practices within a PSM system. And, we have learned that organizational and individual behaviors and culture fuel the engine that implements PSM systems – no matter whether the motivation is for regulatory compliance or simply for good business.

Ensuring that people can return home healthy and uninjured at the end of each workday, ensuring that our neighbors are unharmed, and having a safe work environment have driven many companies to pursue PSM implementation with the objective of having zero incidents. It is that goal for which this guideline was developed – to help companies pursue and achieve the "perfect process safety" vision of zero harm.

1.1 OVERVIEW

It is important to differentiate process safety from other different or broader areas (or management systems) dealing with safety at process plants. For example:

- Process safety is focused on prevention of, preparedness for, mitigation of, response to, and restoration from catastrophic releases of chemicals or energy from an industrial chemical manufacturing process associated with a facility.

- Occupational safety is focused on the prevention of injuries/illnesses to employees due to their tasks or work environment. As such, it tends to focus on hazardous energy related to their personal momentum or the momentum of objects they may be manipulating. Injuries/illnesses could result, such as slips, trips, falls, cuts, thermal burns, musculoskeletal injuries, etc.
- HSE (health, safety, and environment), or the equivalent EHS or SHE acronym, is the broader area that, in addition to process safety and occupational safety, includes occupational health (aka industrial hygiene) and management of environmental impacts.
- SHEQ&S (safety, health, environmental, quality, and security) is the broadest view of related (and hopefully integrated) management systems, as introduced and discussed in *Guidelines for Integrating Management Systems and Metrics to Improve Process Safety Performance* (Ref. 1.1).
- Therefore, process safety is much more than just regulatory compliance (e.g., complying with OSHA's PSM regulation or EPA's risk management program [RMP] rule in the United States).

Historically, most long-established petrochemical companies and facilities (1) started with an initial focus on occupational safety (over 100 years ago in some cases), (2) established occupational health programs as illnesses due to chemical exposures became a known hazard, (3) established environmental programs as public concern increased and regulations were promulgated to protect the environment, and (4) established process safety programs by the 1990s, as guidance and regulations proliferated around the world (see Section 1.2). However, many companies primarily focused their earlier accident prevention efforts on improving their process technology and human factors.

In the mid-1980s, following a series of serious chemical accidents around the world (see Table 1.1 for a summary), companies, industries, and governments began to identify management systems (or the lack thereof) as the underlying cause for these accidents. Companies were already adopting a management systems approach in regard to product quality (e.g., various Total Quality Management initiatives). Companies developed policies, industry groups published standards, and governments issued regulations, all aimed at accelerating the adoption of a management systems approach to process safety. These somewhat fragmented, initial efforts gradually evolved into integrated management systems. The integrated approach remains a very useful way to focus and adopt accident prevention activities. In recent years, inclusion of manufacturing excellence concepts has focused attention on seamless integration of efforts to sustain high levels of performance in manufacturing activities. One goal of manufacturing or operational excellence is to deeply embed PSM practices into a single, well-balanced process for managing manufacturing operations.

Table 1.1 Accidents that Affected PSM Regulatory Development in the USA and Europe

Year	Location	Deaths	Injuries
1974	Flixborough, England	28	?
1976	Seveso, Italy	?	?
1984	Mexico City, Mexico	650	?
1984	Bhopal, India	2,000+	?
1985	Institute, WV	0	135
1988	Norco, LA	5	23
1988	Henderson, NV	2	350
1989	Richmond, CA	0	9
1989	Pasadena, TX	24	132
1990	Channelview, TX	17	0
1990	Cincinnati, OH	2	41
1991	Lake Charles, LA	6	6
1991	Sterlington, LA	8	128
1991	Charleston, SC	9	33

What is a management system? The *Guidelines for Risk Based Process Safety* (Ref. 1.2) define it as:

A formally established and documented set of activities designed to produce specific results in a consistent manner on a sustainable basis.

The *RBPS Guidelines* also emphasize that the management system activities must be defined in sufficient detail for workers to reliably perform the required tasks.

Regarding PSM management systems specifically, the Center for Chemical Process Safety (CCPS) initially compiled a set of important characteristics of a management system, which were published in Appendix A of the *Guidelines for Technical Management of Chemical Process Safety*. Those guidelines were the first generic set of principles to be compiled for use in designing and evaluating process safety management systems. Although Appendix A was groundbreaking, most readers overlooked it as a practical tool because the management systems concept was foreign to them. Since that time, most companies have accumulated significant practical experience in implementing formal process safety, occupational safety, and environmental management systems.

Table 1.2 (originally Table 1.7 in the *RBPS Guidelines*) lists issues that have proven to be most important when designing, developing, installing, revising, operating, evaluating, and improving PSM systems. A PSM framework (such as RBPS) can address one or more of these issues on an element-by-element basis.

The most important thing is that companies thoughtfully consider all of the issues in Table 1.2 when developing a new PSM system, adding new elements, or improving an existing system.

The life cycle of any management system will generally include design, development, rollout, operation, and monitoring/maintenance/improvement. Chapter 4 of these guidelines discusses the overall steps in implementing a new PSM management system:

1. Developing the design specification
2. Creating element and/or system workflows (as appropriate)
3. Estimating element and system workloads and necessary resources
4. Developing the element/system written programs and procedures
5. Rolling out the system
6. Monitoring implementation and initial performance

Similarly, *Guidelines for Integrating Management Systems and Metrics to Improve Process Safety Performance* (Ref. 1.1) discusses the PSM (and overall SHEQ&S) program's life cycles and the Plan-Do-Check-Adjust (PDCA) approach in each chapter. In particular, Chapter 5 discusses how to apply the PDCA approach when implementing a SHEQ&S system, how to set about prioritizing the integration efforts, how to develop integrated systems, and then how to build the concept of continuous improvement into the system's life cycle.

The primary purpose of this book is to provide an update to the original *Guidelines for Implementing Process Safety Management Systems,* recognizing that most companies now have some form of PSM system, but that a number of companies, especially smaller companies or those in developing countries, may need a road map of how to efficiently and effectively upgrade their systems.

Table 1.2 Important Issues to Address in a PSM System

- Purpose and scope
- Personnel roles and responsibilities
- Tasks and procedures
- Necessary input information
- Anticipated results and work products
- Personnel qualifications and training
- Activity triggers, desired schedule, and deadlines
- Necessary resources and tools
- Metrics and continuous improvement
- Management review
- Auditing

1.2 BACKGROUND/HISTORY OF PSM

The American Institute of Chemical Engineers' ' (AIChE's) Center for Chemical Process Safety (CCPS) was established in 1985 as one of the U.S. chemical industry's reactions to a major chemical accident in Bhopal, India. In 1988, CCPS published a motivational advertisement for its forthcoming PSM structure, *Chemical Process Safety Management – A Challenge to Commitment* (Ref. 1.3). This item was intended to educate chief executives in the chemical industry about the importance of implementing PSM activities into their company operations and to motivate them to adopt a management systems approach.

Any discussion on the background and history of PSM would be incomplete without mentioning some other pioneers and pioneering organizations. For example:

Trevor Kletz

After progressing through various positions within Imperial Chemical Industries (ICI), he was appointed as ICI's first Technical Safety Advisor in 1968. During his tenure, ICI developed the hazard and operability (HAZOP) approach and Trevor wrote the first book on this subject. Shortly after retiring in 1982, he expanded an earlier paper entitled "What you don't have, can't leak" into the book that first documented the concept of inherent safety. He is also well known for his many books and presentations emphasizing the importance of learning from previous accidents.

Frank Lees

After working for ICI for a number of years, he joined Loughborough University of Technology and in 1974 was appointed Professor of Plant Engineering. Following the Flixborough disaster that year, he was appointed to the new UK Advisory Committee on Major Hazards. Later, he was a technical assessor for the 1988 Piper Alpha disaster inquiry. He is best remembered for his book *Loss Prevention in the Process Industries*, initially published in two volumes (and over 1,000 pages) in 1980, with the second edition of three volumes published in 1996. (Note that the third edition was published in 2005 by Dr. Sam Mannan and the Mary Kay O'Connor Process Safety Center [discussed below].)

Health and Safety Executive

Shortly after the 1974 Flixborough explosion, the UK promulgated the "Health and Safety at Work" act. This changed the UK approach from one where the authorities defined the procedures for them to follow to one that established goals for operators to meet. Specifically, it replaced the 27 prescriptive acts of parliament with one that transferred the duty for the health and safety of employees and neighbors from the authorities to the employers.

It also established a Health and Safety Executive (HSE) composed of inspectors, specialist scientific, and technical staff to ensure that operators were doing their duty. In order to carry out their responsibilities, inspectors have the authority to enter any facility, take samples, written documents, etc., as they see fit (i.e., without a permit). The HSE and its inspectors follow an enforcement approach that is proportionate to the risks involved, i.e., identifying areas for further improvement through (mandatory) Improvement Notices, Prohibition Notices (to immediately stop operations), and up to prosecutions (for major breaches and/or not following Notices).

The UK implemented the EU's Seveso Directive as the Control of Major Accident Hazard Regulations (COMAH). The HSE reviews documented "Safety Reports," which document the approaches for reducing the risks from Major Accidents Hazards to ALARP (As Low As Reasonably Practicable).

HSE is well known for the technical expertise it demonstrates in regulatory enforcement and development/sharing of guidance documents in this field.

Center for Chemical Process Safety

As discussed in the preface to this book, AIChE created CCPS in 1985 after the chemical disasters in Mexico City, Mexico, and Bhopal, India. The CCPS is chartered to develop and disseminate technical information for use in the prevention of major chemical accidents.

CCPS is a not-for-profit, corporate membership organization within AIChE that identifies and addresses process safety needs within the chemical, pharmaceutical, and petroleum industries. CCPS brings together manufacturers, government agencies, consultants, academia, and insurers to lead the way in improving industrial process safety.

CCPS member companies, working in project subcommittees, define and develop useful, time-tested guidelines that have practical application within industry. The project topics run the gamut of areas important to manufacturers and range from human factor issues to qualitative and quantitative risk analysis to security vulnerability to inherently saferitpm design. With over 100 publications to date, CCPS remains at the forefront of issues relevant to industry.

Mary Kay O'Connor Process Safety Center

The Mary Kay O'Connor Process Safety Center (MKOPSC) at Texas A&M University was established in 1995 in memory of Mary Kay O'Connor, an Operations Superintendent killed in an explosion on October 23, 1989, at the Phillips Petroleum Complex in Pasadena, Texas. Since 1997, the MKOPSC Director has been Dr. Sam Mannan. The Center's mission is to promote safety as second nature in industry around the world in order to prevent future accidents. In addition, the Center develops safer processes, equipment, procedures, and management strategies to minimize losses within the processing industry. It also seeks to advance process safety technologies in order to keep the industry competitive. Finally, the Center (1) seeks to serve all stakeholders (academia, government, industry, and the public), (2) provides a common forum, and (3) develops programs and activities that will forever change the paradigm of process safety. The funding for the Center comes from a combination of an endowment, consortium funding, and contract projects.

Also, see several articles in the June 2009 edition of *Process Safety Progress* (Ref. 1.4) for additional information on the history of process safety.

1.3 PROCESS SAFETY RESOURCES

In 1989, CCPS began publishing a series of guidelines, starting with *Guidelines for Technical Management of Chemical Process Safety*, to encourage its members to pursue accident prevention in more integrated, holistic ways.

In 2007, CCPS published *Guidelines for Risk Based Process Safety*, which laid out the next generation, 20-element management system for process safety. In total, the CCPS has published more than 100 guidelines, tools, and concept books covering a wide range of PSM-related topics. Table 1.3 lists some of the key guidelines and tools that have paved the way for companies seeking to adopt, implement, and improve PSM management systems.

In addition, Appendix III of this book provides an extensive listing of RBPS implementation tools, along with summaries of the purpose of each tool and examples of many of the tools (typically, by references to Web sites or to the files on the Web accompanying this book).

Other industry groups and government agencies also developed PSM frameworks, and Tables 1.4 and 1.5 list a sampling of these. Most of the frameworks are similar in construction, include identical or similar safety management system elements, and promote similar process safety work activities. However, differences exist in the frameworks, particularly the newer ones. In many cases, the sponsoring country or organization wisely looked around the world and then built its process safety structure based on current best practices within the industry.

In summary, PSM has advanced and today there are many process safety models, support tools, and organizations available to help advance process safety and how organizations and individuals stay engaged and involved (i.e., promote continuous education and innovation). Process safety successes and failures depend upon dedicated knowledgeable individuals throughout our industry, governments, and academia working together toward the common goal of preventing catastrophic incidents.

Table 1.3 CCPS Guidelines and Tools for Chemical Process Safety Management

- *Guidelines for Technical Management of Chemical Process Safety*, 1989
- *Plant Guidelines for Technical Management of Chemical Process Safety*, 1992, 1995
- *Guidelines for Hazard Evaluation Procedures*, 1992, 2008
- *Guidelines for Investigating Chemical Process Incidents*, 1992, 2003
- *Guidelines for Auditing Process Safety Management Systems*, 1993, 2011
- *Emergency Relief System Design Using DIERS Technology*, 1993
- *Guidelines for Safe Automation of Chemical Processes*, 1993
- *Guidelines for Implementing Process Safety Management Systems*, 1994
- *Guidelines for Integrating Process Safety Management, Environment, Safety, Health and Quality*, 1996
- *Guidelines for Writing Effective Operating and Maintenance Procedures*, 1996
- *Guidelines for Pressure Relief and Effluent Handling Systems*, 1998
- *ProSmart: Performance Measurement of Process Safety Management Systems*, 2001
- *Layer of Protection Analysis: Simplified Process Risk Assessment*, 2001
- *Guidelines for Mechanical Integrity Systems*, 2006
- *Guidelines for Risk Based Process Safety*, 2007
- *Guidelines for Performing Effective Pre-Startup Safety Reviews*, 2007
- *Guidelines for Safe and Reliable Instrumented Protective Systems*, 2007
- *Guidelines for the Management of Change for Process Safety*, 2008
- *Guidelines for Process Safety Metrics*, 2009
- *Guidelines for Evaluating Process Plant Buildings for External Explosions, Fires, and Toxic Releases, 2nd Edition*, 2012
- *Guidelines for Engineering Design for Process Safety, 2nd Edition*, 2012
- *Guidelines for Enabling Conditions and Conditional Modifiers in Layers of Protection Analysis*, 2013
- *Guidelines for Integrating Management Systems and Metrics to Improve Process Safety Performance*, 2015

Table 1.4 Significant Industry-Based PSM Initiatives

- Chemistry Industry Association of Canada (formerly Canadian Chemical Producers Association): program, 1986
- American Chemistry Council (formerly Chemical Manufacturers Association): Responsible Care Initiative Process Safety Code of Management Practices, 1987, 2013
- AIChE Center for Chemical Process Safety: Technical Management of Chemical Process Safety, 1989
- American Petroleum Institute Recommended Practice 750 – Management of Process Hazards, 1990
- ISO 14001: 1996 and 2001 – Environmental Management System
- Organization for Economic Cooperation and Development Guiding Principles on Chemical Accident Prevention, Preparedness, and Response, 2003
- American Chemistry Council Responsible Care® Management Systems and RC 14001, 2004
- [UK] Energy Institute: High Level Framework for Process Safety Management, 2010
- Canadian Society for Chemical Engineering: Process Safety Management Standard and Guide, 2012
- The American Fuel and Petrochemicals Manufactures and American Petroleum Institute's "Advancing Process Safety" initiative. (Programs include process safety metrics, event sharing, process safety hazards identification, process safety regional networks, and process safety site assessments.) See www.afpm.org/policy-position-process-safety/ for more information.

Some of these PSM frameworks are discussed in more detail in Chapter 4.

Table 1.5 Partial List of Worldwide Governmental Accident Prevention and PSM Initiatives

- European Commission: Seveso I Directive, 1982; Seveso II Directive, 1997; Seveso III Directive, 2012
- U.S. Occupational Safety and Health Administration Process Safety Management of Highly Hazardous Chemicals (29 CFR 1910.119, 1992
- U.S. Clean Air Act Amendments: Section 112(r) – Accident Prevention, 1992
- U.S. Environmental Protection Agency Risk Management Program rule (40 CFR 68, 1996
- Mexico: Integral Security and Environmental Management System (SIASPA), 1998
- United Kingdom: Health and Safety Executive COMAH regulations – The Control of Major Accident Hazards Regulations, 1999 (amended in 2005 and 2015)
- Australia: Occupational Health and Safety Act 1985 Occupational Health and Safety (Major Hazard Facilities) Regulations 1999 (SR 1999). National Standard for the Control of Major Hazard Facilities [NOHSC1014(1996/2002)]. Work Health and Safety, 2011
- Canada: Canadian Environmental Protection Act – Environmental Emergency Regulation, Section 200 Part 8, 1999
- Republic of Korea: Korean OSHA PSM standard, Industrial Safety and Health Act – Article 20, Preparation of Safety and Health Management Regulations. Korean Ministry of Environment – Framework Plan on Hazardous Chemicals Management, 2001-2005
- Japan: High Pressure Gas Safety Act, 2006
- Brazil: ANP Oil and Gas industry accident prevention regulations
- Malaysia: Department of Occupational Safety and Health, Ministry of Human Resources, Section 16 of Act 514
- Singapore Standard SS506 Part 3: Occupational Safety and Health (OSH) Management System – Requirements for the Chemical Industry, 2013
- China: Guidelines for Process Safety for Petrochemical Corporations – AQ/T3034, 2010
- U.S. Bureau of Safety and Environmental Enforcement: Safety and Environmental Management Systems, 2011
- International Association of Oil and Gas Producers: Process Safety – Recommended Practice on Key Performance Indicators, 2011
- Mexico: NOM-028-STPS-2004, Process Safety and Critical Equipment Handling Hazardous Chemicals System, 2012
- European Union: EU Directive 2013/30/EU on Safety of Offshore Oil and Gas Operations, 2013

See Appendix I for a complete listing and additional information on these regulations/initiatives.

1.4 PSM IMPLEMENTATION LESSONS

Various factors can continuously or periodically influence a company's PSM system implementation and/or performance; examples include:

- Significant internal or external incidents, which point out actual or potential weaknesses or new areas that need to be addressed
- Economic conditions, which may bring pressure to reduce the costs and resources associated with maintaining systems
- Process changes or mergers/acquisitions that introduce new processes/ chemicals with new hazards and risks. For example, a small site may not have previously been required to implement a PSM system (due to either regulatory or corporate requirements), but now:
    - it increases the quantity of a highly hazardous chemical used in the process and now needs a formal PSM system that will ensure a higher level of attention to process safety, or
    - it is acquired by a different company that requires a formal PSM system to be instituted due to the chemicals/quantities handled in the process, to reduce the risk to employees and neighbors, etc.
- Workforce shifts, where experienced PSM personnel leave or move to different roles, resulting in a reduction of knowledge/experience
- Organizational changes, which either leave some key PSM system responsibilities unassigned or move experienced PSM personnel to different roles
- Hiring of new college graduates with engineering and other professional technical majors but without ensuring adequate PSM training and education for them prior to their involvement in PSM processes
- Regulatory changes, which add new requirements that the PSM system must address
- Global expansion, leading to issues such as maintaining the PSM system robustness and fitness-for-purpose as the company gets larger, integrating the PSM system of a new acquisition, and instilling the desired safety culture in personnel in various countries

These and other influences may lead to companies seeking new ways to improve PSM system activities based on strategies such as the following:

- Decreasing or eliminating PSM system activities that are judged as overly demanding or unnecessary, based on risk judgments
- Performing PSM system activities more efficiently
- Using the same resources, but using better practices to generate improved results
- Getting better PSM results, but with fewer resources

- Extending existing PSM system practices and activities into new areas
- Extending existing PSM practices throughout the management system life cycle (e.g., an Operational Excellence approach
- Adding new PSM activities to existing PSM elements
- Creating new PSM elements
- Restructuring the PSM system
- Establishing in-house PSM training curriculums for employees at all levels

In the last 25 years during which PSM systems have become more and more common, many lessons – both positive and negative – about PSM implementation have been learned. Some examples of these lessons are briefly discussed or referenced in the various chapters in this book, but in general the positive lessons include factors such as:

- good planning,
- adequate and knowledgeable resources, and
- continuous learning and improvement/innovation.

In addition, the appendices of this book and/or the files on the Web accompanying it include a number of PSM implementation lessons, as well as PSM implementation resources, including the following:

- A case study of Eli Lilly and Company's PSM implementation experience (Appendix II)
- A number of PSM system tools/resources shared by Eli Lilly and Company (on the Web)
- An extensive list of "RBPS Implementation" tools (Appendix III)
- A description of how to map PSM system/element performance issues to culture features (Appendix XIV)
- An example of a Process Safety Culture Survey (on the Web)
- A detailed PSM project implementation plan example
- A current compilation of PSM-related software
- A set of contractor safety and health guidelines

1.5 THE BUSINESS CASE FOR PROCESS SAFETY

As process safety became more and more common for companies and sites during the 1990s, process safety professionals found that they were often asked – and asked themselves – one question: What is the business benefit for process safety?

The easiest answer to this question comes from the costs of a lack of proper process safety management, i.e. process safety events. The Marsh 100 Largest Losses (1974-2013) estimated the total cost of property damage over this period to be $34 billion. These accidents "generally occur because of the failure of a number of the systems or barriers within the process-safety management systems." The $34 billion figure is for property damage alone. It ignores the fatalities that result and the additional costs to companies and society from the incident; for example: (1) Bhopal (over 2,000 fatalities and $400 million), (2) Flixborough (28 fatalities), (3) Buncefield (£1 billion), (4) Longford (two fatalities and $1.3 billion), and (5) Macondo (12 fatalities and over $30 billion).

In an effort to answer this question and show the business benefits from a strong PSM program, CCPS commissioned a study and developed an initial brochure on "The Business Case for Process Safety" in 2006, which was subsequently upgraded and revised in 2010 (available in Appendix IV). In addition, Project 245 ("Business Case for Process Safety and Sustainability") intends to update the original material with current examples and expand it to include the concept of sustainability.

The study identified two qualitative and two quantitative benefits for process safety:

- Qualitative benefits:
  - Corporate responsibility – process safety protects a company's image, reputation, and brand.
  - Business flexibility – process safety preserves a company's license to operate and gives it increased business options.
- Quantitative benefits:
  - Risk reduction – process safety prevents human injury and avoids significant losses and environmental damage.
  - Sustained value – process safety helps boosts productivity and produce high-quality products, on time and at lower cost, which contributes to shareholder value.

In terms of real, measurable benefits, the companies that participated in this study reported significant direct cost benefits of up to:

- 5% increase in productivity,
- 3% reduction in production costs,
- 5% reduction in maintenance costs,
- 1% reduction in capital budget, and
- 20% reduction in insurance costs.

In order to realize these benefits, the study recommends seven steps for achieving business excellence through process safety management:

1. Assign personnel who will be accountable. Typically, either a process safety manager or team should be responsible for (a) ensuring excellence in pursuing process safety throughout the corporation, (b) reevaluating your program's effectiveness, (c) estimating your company's and sites' "process safety return on investment," and (d) communicating it to the employees and the public.
2. Adopt a personalized company philosophy of process safety. Use it to establish a management system along the lines of CCPS guidelines (referenced in this book) and tie it into your company's core values.
3. Learn more about process safety by reviewing the literature and other references, attending training provided by process safety professionals, and interacting with other companies (e.g., networking with them and participating in industry alliances).
4. Take advantage of the strong synergy process safety has with your other business drivers. For example, Total Quality Management (TQM), regulatory requirements, and the American Chemistry Council's (ACC's) Responsible Care® initiative all share common elements.
5. Set achievable process safety goals that will support the business case presented over the next one to five years.
6. Track your performance versus goals periodically (note that this book stresses the importance of monitoring and metrics, and provides references on these subjects).
7. Revisit your process safety program and modify it every three to five years as needed. (Clearly, this book is intended to help guide any PSM system modification or upgrade efforts.)

Keeping in mind the importance of making the business case for PSM periodically within your site/company, it is a good idea to continuously look for and capture PSM implementation benefits as your organization continues its PSM journey.

1.6 IMPORTANCE OF INTEGRATING PSM WITH BUSINESS SYSTEMS

While PSM systems can stand alone, PSM systems reach far beyond process safety objectives and results. PSM systems are well aligned with business systems and achieve business objectives and results, along with process safety risk reduction. Examples include management systems for the following:

- Process safety information (PSI). PSI management systems often go beyond PSI and approach intellectual property (IP) or other technical knowledge.

- Process hazard analyses (PHAs). Companies often use PHAs to go beyond process hazards and also analyze business risks.
- Operating procedures for PSM as well as procedures for business processes
- Contractor management for process safety as well as for business processes
- Mechanical integrity (MI). MI can be extended to increase equipment reliability and plant uptime.
- Incident investigation techniques, which can be applied to "loss of production" and equipment failure incidents
- Management of change (MOC). Similar change management rigor can be applied to business processes.

"PSM systems" are rarely just PSM systems. Their objectives and results go beyond process safety and generally create reliable sustained operations. PSM systems may be looked at as specific or "focused" business systems and should be integrated into the company's business systems and practices at every level. Specific process safety objectives and results should be documented, highlighted, and understood for each business system as well as other objectives and required results.

PSM systems tend to share some common management system needs (e.g., planning, budgeting, training, risk analysis, change management, off-normal event reporting and investigation/analysis, contingency planning, auditing, performance analysis, management review) with other management systems. For many companies, it makes sense to standardize key aspects of these common/similar elements. The more integrated a PSM system is with either the HSE system or the business management system (BMS), the greater the likelihood that the promise of consistency and efficiency can be achieved. (Note that the recent CCPS book entitled *Integrating Management Systems and Metrics to Improve Process Safety Performance* [Ref. 1.1] focuses on this important topic and provides extensive guidance on this subject.)

In addition, most companies face overlapping regulatory, industry and trade association, and certification requirements that can consume significant resources and attention. Combining the synergies among these various business systems will help ensure safe and reliable operations, streamline procedures and cross-system auditing, and support regulatory and corporate compliance requirements. Since some of the systems are common to more than one area, a well-designed and well-implemented integrated management system will help reduce the load on the process safety and other groups. In addition, an integrated system will help improve manufacturing efficiency and customer satisfaction. Further, the importance of integrating process safety, health, environmental, quality, and security performance improvement systems has been noted in recent conferences, webinars, journals, and books.

Whether a facility is regulated or not, if it must handle hazardous materials, a company's success will be favorably impacted when it applies the fundamental elements of a PSM program within its business systems and other risk reduction programs. In addition to regulations, societal and political pressures from the public demand ever-better safety and environmental performance. So, every company needs to find ways to improve its operating efficiency and performance, reduce overall operating cost, and at the same time find ways to maintain and improve its competitive market position.

Although the management programs for process safety and other business systems may have been developed separately, they have similar program-related expectations, such as:

- a formal, implemented program;
- specific program-related recordkeeping requirements; and
- metrics used to demonstrate performance program improvements.

Due to the different, sometimes conflicting goals for each group, the demands on an operating facility may inadvertently prompt unsafe program changes and contribute to an increased process safety-related operating risk. A formalized, integrated, and well-managed system helps provide the controls that prevent such changes from occurring.

The potential high-level benefits of integrating PSM with other business management systems include lower costs, improved problem solving, work process consistency, continuous improvements, clearly identified measures, sound statistical data analyses, and satisfied and engaged customers (Ref. 1.1). Other benefits of integrating PSM into business systems are those discussed in Section 1.4 ("The Business Case for Process Safety") of these guidelines and in the CCPS brochure provided in Appendix IV.

In summary, there are many benefits to integrating PSM into business systems, and doing so is vital to successful PSM system. The most successful companies will be the companies that integrate process safety into their business systems and practices, understanding how each business system impacts process safety and highlighting it to ensure that process safety is sustained over the life cycle.

1.7 INTENDED AUDIENCE AND HOW TO USE THESE GUIDELINES

These guidelines are intended for use by facility or corporate personnel responsible for designing, implementing, or monitoring the performance of PSM systems for facilities. Typical facility personnel job roles would include plant engineers or technical specialists involved with executing specific PSM element activities, element coordinators, and PSM/HSE managers. Typical corporate personnel would include PSM element subject matter experts and PSM/HSE managers.

In addition, anyone who is in a position to evaluate, plan, coordinate, advise, or execute PSM/HSE implementation, integration, or improvement efforts may benefit from these guidelines; for example:

- Corporate PSM/HSE coordinators
- Corporate PSM element subject matter experts
- Facility/asset PSM/HSE managers and coordinators
- PSM/HSE element champions and subject matter experts
- Plant engineers
- Engineering and construction firms
- PSM/HSE consultants

Companies can use the information provided in this book to help perform one or more of the following tasks:

- Determine process safety implementation and performance status
- Prepare for PSM system change
- Implement a new PSM system
- Incorporate new elements into an existing PSM system
- Improve an existing PSM element or system
- Integrate PSM/HSE with a business management system
- Manage future process safety performance

This book devotes chapters to each of these PSM activities. Personnel involved in any of them can consider the features described for each activity. Several appendices provide additional information useful to those personnel.

Table 1.6 lists perceived user needs and provides guidance on how to use this book to best meet those needs.

Table 1.6 Roadmap for Using This Book to Implement PSM

User Need Description	Contents to Review to Meet Needs
Want to know the basics	1
Evaluate PSM implementation and performance	1, 2
Want to prepare the organization for the change	1, 3, Appendix VII
Develop and/or implement a new PSM system	1, 4, Appendices II and III
Add new elements to an existing PSM system	1, 5, Appendices II and III
Improve an existing PSM element or system	1, 2, 6, Appendix III
Integrate PSM with other business systems	1, 7, Appendix IV
Sustain or improve PSM performance	1, 8

1.8 REFERENCES

1.1 Center for Chemical Process Safety of the American Institute of Chemical Engineers, *Guidelines for Integrating Management Systems and Metrics to Improve Process Safety Performance*, John Wiley & Sons, Inc., Hoboken, New Jersey, 2015.

1.2 Center for Chemical Process Safety of the American Institute of Chemical Engineers, *Guidelines for Risk Based Process Safety*, John Wiley & Sons, Inc., Hoboken, New Jersey, 2007.

1.3 Center for Chemical Process Safety of the American Institute of Chemical Engineers, *Chemical Process Safety Management – A Challenge to Commitment*, New York, New York, 1988.

1.4 American Institute of Chemical Engineers, "History of Process Safety" (several articles), *Process Safety Progress*, New York, New York, Vol. 28, Issue 2, June 2009, pp. 103-207.

2

EVALUATING PSM SYSTEM IMPLEMENTATION AND PERFORMANCE

A vital part of any sustainable management system is monitoring the system status and performance and making adjustments as needed. Determining the status is important in order to (1) know where to start, (2) understand how much effort will be required, (3) identify opportunities that will provide the best return on time and investment, and (4) perform the "check" portion of the PDCA life cycle.

The status of a facility's PSM system implementation and its actual process safety performance depend on many interconnected factors. The techniques used to evaluate implementation and performance at a large petrochemical complex will differ from those used at a small, remote, single-unit facility, or within a multinational organization that operates several different types of facilities worldwide in different operating sectors (e.g., upstream exploration and production and downstream refining and distribution). Among the many factors that influence the status of process safety implementation and performance are (1) the facility's size and age, (2) the severity of consequences associated with process hazards, (3) the safety culture of employees and managers, (4) significant incidents at this and other facilities within the company, and (5) the regulatory requirements and climate in the area where the facility is located. Furthermore, the techniques (or tools) used to evaluate implementation status and performance will vary depending on the life-cycle stage of the facility. Therefore, to determine the status at the facility, the right techniques should be selected based on its life-cycle stage as well as other facility-specific factors.

2.1 THE MODIFIED SAFETY TRIANGLE

The safety triangle (or pyramid) has frequently been used to illustrate that accidents do not occur in isolation, but are instead the result of failures of underlying systems or precursors. The triangle has been modified to include management system failures as a contributor. Subsequently, unsafe behaviors and attitudes and the safety culture were added to illustrate even more deep-seated aspects that, if not adequately understood and managed, can ultimately lead to more serious events.

The safety culture is addressed to some extent in the "attitudes" descriptor, but the safety culture as a whole is more than just the attitudes of the workers; it is a reflection of the safety culture throughout the organization (see Figure 2.1).

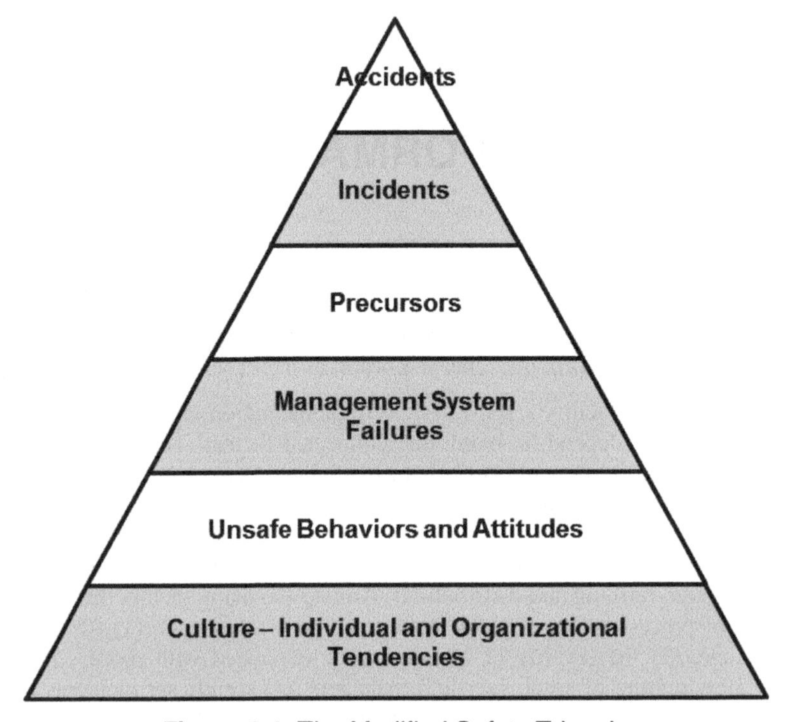

Figure 2.1 The Modified Safety Triangle

To get an accurate picture of the process safety program at a facility, it is important to examine and measure performance of the underlying systems as well as process safety precursors, incidents, and accidents at each level. Further, drawing accurate conclusions about the status of the process safety program as a whole requires a detailed review at all levels. By looking holistically up and down the triangle, the strengths of a program can be identified and shared with other areas, and weaknesses can be identified as the area(s) where applying more resources will have the largest impact on improving process safety performance.

Note that the levels in the triangle are set up to differentiate measurements; however, PSM and other management systems can cut across multiple levels of the triangle. To some extent a management system should drive behaviors and attitudes and reflect the organizational culture and company/site risk profile, as well as drive how system failures (i.e., precursors, incidents, and accidents) are handled. Our examinations and measurements should verify that:

- our cultures, individual and organizational, are reflected in our systems;
- our behaviors are driven by the system requirements;
- system failures are identified, measured, and acted upon; and
- process safety precursors, incidents, and accidents are well understood by personnel and shared with the site, company, and/or industry as appropriate.

2.2 COMMON INDICATORS AT EACH LEVEL OF THE TRIANGLE

There are process safety indicators (metrics or key performance indicators [KPIs]), that can be tracked at every level of the triangle. The type and number of metrics that should be tracked vary from company to company and facility to facility, and even from process to process within a facility. The types of metrics to be tracked should be kept evergreen and subject to periodic review and adjustment. The metrics should be a mixture of leading and lagging indicators and should consider, at a minimum, the following:

- Historical trends in process safety performance
- Recent incidents
- Corporate/company/site emphasis programs
- Audit results
- Metrics as required by regulatory requirements
- Weaknesses as revealed in management reviews and process safety evaluations

This book does not go into detail on how to conduct an effective PSM-related audit or how to effectively use the results of an audit for continuous improvement. Instead, the reader is referred to the CCPS book entitled *Guidelines for Auditing Process Safety Management Systems, 2nd Edition* (Ref. 2.1), for helpful details on this topic.

Following are examples of metrics that can be considered at each level:

- **Accidents.** The number of process safety incidents, severity of the incidents (extent of harm to people, the environment, assets, etc.), common breakdowns or contributing factors and causes
- **Incidents.** The number of first-aid incidents, severity rate, small releases and fires, number of emergency response team callouts, common breakdowns or contributing factors and causes
- **Precursors.** The number of process safety near misses reported, number of unsafe conditions reported, number of demands on safety systems, common breakdowns or contributing factors and causes

- **Management system failures.** HSE audit score and findings, number of overdue action items, corrective actions generated, safety meeting attendance, training completed, overdue/delayed MI inspections, alarm frequency, evaluation of incident investigation effectiveness
- **Unsafe behaviors and attitudes.** Safety inspections completed, behavior-based safety observations completed
- **Culture – individual and organizational tendencies.** Survey scores, interviews, management visits to the processes, housekeeping, evaluation of communication frequency and effectiveness

This book does not detail how best to establish meaningful metrics, but resources from CCPS, the American National Standards Institute (ANSI), and the American Petroleum Institute (API) are available for use by a facility wanting to implement or upgrade its process safety metrics (Refs. 2.2 and 2.3).

2.3 PROCESS STAGES IN THE COMPANY/FACILITY LIFE CYCLE

The parameters that can be reviewed to determine the degree of PSM system or program implementation and the process safety performance status of that program vary based on the life-cycle stage of the company and/or process. For example, in the conceptual and early stages of a process, there is limited (or no) data available for that specific process. In these cases, indicating data cannot be drawn from the traditional sources such as auditing and incident investigations. However, even in this example of a facility in its early stages, process safety issues should have been considered in the design. For example, adequate equipment spacing and pad drainage would have been key components in the development of the plot plan. A careful review of the design through the "lens" of process safety can be a strong contributor to successful process safety performance when the process is running. Furthermore, some information is available from similar processes both inside and outside of this company. The judgment of experienced operators, maintenance technicians, and engineers can also be valuable in establishing the programs.

As with all aspects of safe operation of a process, "buy-in" by facility management is the key to the successful implementation of a process safety program leading to strong process safety performance. The cooperation of facility-based staff and managers can strongly influence the effectiveness of baseline PSM assessments. The assessment phase may be the first visible manifestation of your company's PSM initiative at the local level, and it should be properly understood by those participating – especially those whose PSM activities and programs are under review. An assessment method with which local personnel are at least somewhat familiar, conducted by professionals whose skills they respect, will more likely gain buy-in and cooperation.

For this discussion, the company/facility life cycle is divided into the following five stages:

1. Facility in design phases (prior to startup)
2. New company/facility (shortly after forming or startup)
3. Acquisition of company/facility
4. Facility expansion
5. Existing facility

Techniques that can be used to assess the status of PSM system implementation and process safety performance at the various life-cycle stages will include the following:

- An evaluation and summation of audit results (e.g., gap analysis, regulatory coverage assessments, recognized and generally accepted good engineering practice [RAGAGEP] assessments)
- A review of incident and near-miss incident root cause reports to look for trends and common weaknesses. The results of this review should be compared with PHAs or other risk reviews to help ensure that the hazards from actual events are included in the risk review documentation, and that the corrective actions are incorporated into safeguards, where appropriate.
- A summation of the data collected from lagging indicator metrics (e.g., accidents, incidents). Similar to the review of incidents and near misses, the data from lagging indicators should be reviewed with the goal of continuous improvement, looking for trends and common weaknesses to be corrected and proper documentation in PHAs and hazard reviews.
- The development of leading indicators and incorporation of the metrics from those indicators in order to (1) improve the process safety program specifically based on the needs of the facility, (2) drive improved process safety performance due to setting specific goals (or metrics) and by requiring periodic reporting against those metrics, and (3) maintain open lines of communication with all facility personnel regarding process safety performance against the metrics
- The development and use of other performance characteristics (e.g., safety culture studies, employee surveys, input from the community via community advisory panels)

The value of the technique will vary based on the life-cycle stage and the available data.

No matter which stage your company/facility is in, planning is the key, and incorporating process safety goals, objectives, and milestones is imperative. Identifying resources with the knowledge and experience to represent process safety interests will ensure process safety performance. “Fail to plan; plan to fail.”

Stage 1: facility in design phases (prior to startup)

The facility has been conceived on paper and may even have been constructed, but not started up. The process safety program has to be designed and constructed in the same way the equipment is designed and constructed.

PSM program implementation. The program does not yet exist and therefore has not been implemented. This is a great opportunity to take a clean sheet of paper and design the type of process safety program that will be most effective for this facility. Key initial steps will generally include the following:

1. Identify knowledgeable resources who have experience with implementing and integrating PSM systems. It is important to identify and involve resources knowledgeable and experienced in process safety as early as possible and to include them in every phase through startup and steady state. They should ensure that process safety is fully integrated into the planning and appropriate systems as well as begin to define the culture.
2. With the help of those resources, develop plans that incorporate process safety goals, objectives, and milestones at each phase.
3. Develop and implement PSM management systems with clearly defined strategic objectives that include and highlight process safety objectives.

The employee participation plan required under the PSM standard is a good starting point for organizing the program. By “beginning with the end in mind” (Ref. 2.4) when developing the PSM management system, all necessary aspects are covered, accountabilities are assigned, and mechanisms are in place for continuous improvement.

Develop the new program based on internal standards, regulatory requirements, and/or the established programs at other company facilities. Refer to the guidelines in Chapter 4 of this book. Facilities that are covered under OSHA’s PSM standard will have structured requirements under the 14 elements given in the standard to help provide structure. Existing programs in place at other units within the company, especially those with similar technologies, materials, or hazards, would be the logical resources with which to begin the program and systems development.

Process safety performance. No specific historical data are available. Initial performance targets can be set shortly before startup based on company standards, similar processes, or industry experience with this or similar processes. Experienced personnel can (and should) be used to set the initial metrics based on the collective pool of experience. Baseline data and metrics should be collected upon startup and used to drive process safety performance improvements. In general, following startup data and metrics should be broader to determine and verify that all systems are functioning as needed. Upon verification of systems, metrics should be adjusted and rotated as needed to align with performance and business needs (consult available resources for how best to use measures to drive performance).

Stage 2: new company/facility (shortly after forming or startup)

PSM program implementation. If there is no program in place, then the approach would generally be the same as described above for a facility in the design phases. However, there are two key differences:

1. The facility should take a risk based implementation approach that assesses the highest risks first and evaluates whether any immediate operational changes are needed, including shutting down until the appropriate PSM systems can be implemented to ensure that the company's risk profile is maintained.
2. There is typically a flurry of activity surrounding a new company or facility startup with organizational changes, process changes, operating procedure revisions, and the normal troubleshooting and correction that may be required from the same personnel who would typically be tasked to develop the PSM programs. The push for competing resources makes this stage perhaps the least desirable for program development.

On the other hand, if the program is in place, albeit fairly new and untested, facility personnel should conduct an initial gap assessment as soon as practical. Consider doing this soon after the process achieves steady state using a PSM or RBPS audit protocol that includes all applicable elements. Detailed audit guidance can be found in *Guidelines for Auditing Process Safety Management Systems* (Ref. 2.1). Waiting to do this gap assessment during the first PSM compliance audit puts the facility in a reactive mode. Latent hazards can be undetected until discovered through an audit. It can also increase process safety and regulatory risks for up to three years prior to conducting the first formal compliance audit.

Process safety performance. Limited data are available, so actual performance trends cannot be accurately assessed. However, any incidents, near misses, or hazardous conditions identified during startup or in the early stages of operation should be captured as institutional knowledge that may apply to other startups, shutdowns, and perhaps during normal operation. New facilities could also consider looking externally at other similar facilities (or industries) to determine whether there are any appropriate lessons learned to be taken into account. Ensure that this information is included in the PHA documentation as PSI to be used in subsequent PHAs. This information should also be communicated to all affected personnel and incorporated into training and procedures, as appropriate.

Early performance information can also be used to adjust initial targets and metrics as data become available (consult available resources for how best to use measures to drive performance).

Stage 3: acquisition of a company/facility

When a facility or company is acquired, assessing the degree of PSM program implementation should be one of the key initial, post-acquisition activities.

PSM program implementation. Assessing PSM program implementation and performance status will be one of the key activities in the due diligence process. However, by its nature due diligence covers many topics, and the effort that can be invested in the PSM-related review typically does not provide the depth and breadth needed for a truly accurate picture. Therefore, it is suggested that the new owners conduct a full, detailed gap assessment as soon as possible after the acquisition is finalized using a PSM or RBPS audit protocol that includes all applicable elements (Ref. 2.5). Focus on the most recent PSM compliance audit and identify management systems that have been used to ensure continuous compliance.

The integration plan should include an initial assessment that evaluates PSM and RBPS, including risk profile and culture differences. The output of the assessment should be to update the plan to close gaps based on the associated risks and include the resources needed.

The new owners should execute a thorough facility siting and consequence analysis study to understand the facility's vulnerabilities, especially those related to the hazards to personnel in occupied buildings. Following the audits and reviews, they should establish a detailed, realistic (but aggressive) action plan to close the gaps. Develop management systems where necessary and update the program as needed based on the audits and reviews.

Facility management should also consider initiating a project to assess the facility process safety culture and take corrective action for process safety culture improvement identified in the assessment, as appropriate for the facility. It is a near certainty that the safety cultures of the two merged companies will differ. Sometimes the difference is dramatic. By evaluating the safety culture, steps can be taken to unify the safety message and begin forming a unified safety culture. A well-conducted safety culture study can reveal attitudes that, if not corrected, can lead over time to less-than-satisfactory safety performance results.

Process safety performance. A better understanding of the process safety performance at the facility can result from compiling incident investigation and root cause information data from all sources and looking for trends and common/repeat failures. Additional information can be gained by evaluating existing lagging metrics and analyzing the trends. The results should be communicated and displayed; the analysis of performance trends can then be applied to the current facility.

Facility personnel should also evaluate existing leading metrics for completeness and relevance to the current facility condition/situation. The results should be communicated and displayed; the analysis of performance trends can then be applied to the current facility.

Stage 4: facility expansion

A significant facility expansion presents unique challenges to the process safety programs. Expansion projects are done (obviously) to increase the unit capacity or otherwise respond to market conditions. Facilities can be modified to implement process safety improvements (e.g., new safety instrumented system), but they are not expanded with process safety as the key design variable. It is therefore imperative that the process safety function be represented on the expansion project team from the start and included in each phase of the plan and execution. The process safety personnel help the other expansion project personnel understand how process safety is affected by the expansion. For example, PSM elements such as (1) process safety information, (2) process hazard analysis, (3) operating procedures, (4) training, (5) mechanical integrity, (6) management of change, and (7) pre-startup safety review (and others) will be directly impacted by the change. The process safety personnel will ensure that the plan includes the appropriate process safety requirements at each phase and verify the quality of development and implementation of the systems.

One trap to be especially aware of is the common notion that the new equipment is "exactly" like another part of the facility – built as a "cookie cutter" facility. This is a myth. There are never two processes or parts of the process that are exactly the same. Even if they are built in an identical manner, they are not located in the exact same spot; so facility siting questions, for example, need to be examined. Further, the second facility design is often modified due to lessons learned from the first facility (e.g., materials of construction, instrumentation). Finally, even if the facilities are identical, they won't be operated by the same people. These differences can fundamentally affect process safety.

With the changes inherent to a large-scale expansion project, it is therefore critical that all PSM aspects of the expansion be carefully reviewed, just as though it is a new facility. The difference, of course, is that the facility has the advantage of incorporating existing programs (e.g., inspection, testing, and preventive maintenance [ITPM]; procedures; training) and using the knowledge gained from previous audits, inspections, incidents, and near misses as starting points for the expanded process. Taking such lessons learned into account helps avoid repetition of previous incidents.

PSM program implementation. The evaluation strategy for the PSM program status of an expanded facility will fall between the new facility stage (described above) and the existing facility stage (described below). The same evaluation tools applicable to regulatory and internal requirements used for a new facility will be used to capture impacts that the new expansion will have on the existing process. The current, well-run programs from the existing facility should be applied to the expanded facility.

The facility expansion presents a very good opportunity for fine-tuning the existing program by building on strengths (those portions of the PSM program that are clear successes) and modifying/improving areas that are not working well. For example, if the PHA program is considered to be thorough and effective, the facility personnel may consider integrating the knowledge gained from the PHA to develop a more robust and consistent MI program. Namely, the equipment and other instruments identified as safeguards in the PHA should (1) be consistently identified as critical equipment, (2) be included in the ITPM plan, and (3) serve as the basis for the safe operating limits documentation.

Process safety performance. While there will be no initial performance data for the expanded facility, there will be ample data from the original facility. The existing metrics, goals, and targets will be the starting point for evaluating performance, and these should be reviewed and modified as needed to account for the expanded facility.

When evaluating PSM program implementation and process safety performance in an expanded facility, the precaution is not to assume that existing programs are adequate without careful review. By virtue of the expansion, the unit may be dramatically different. Due to different flow rates, capacities, equipment sizes, compositions, and other process parameters, new hazards and safeguards may be introduced with the expansion. Additional (or revised) PSI will be required. Facility personnel should review the expanded process taking into account all 14 OSHA PSM elements (and 20 RBPS elements) to fully appreciate the changes affecting the PSM program and potentially process safety performance.

Stage 5: Existing facility

The existing facility, typically with years of operating history, will have a PSM program that can be evaluated to ensure that it has all the necessary parts and pieces, and the results and status by which it can be measured.

PSM program implementation. Periodic compliance audits of the existing PSM program are a regulatory necessity and a practical imperative. You will only improve what you measure. What may be overlooked is a periodic audit of the management system. Actually, this inward evaluation should be an important and visible element of the management system itself. A periodic evaluation of the PSM program, with mid-course corrections, will help ensure a vibrant and evergreen program. Important related considerations include (1) ensuring that evaluations are part of the organization's periodic (e.g., at least annual) planning process, (2) identifying focus areas based on analyses and identified risks, and (3) ensuring that HSE (including process safety) is identified, highlighted, and incorporated into every planned initiative, as appropriate.

Having an effective means to track and ensure completion of corrective actions from all program review activities is important and also a regulatory necessity. Establish a detailed, realistic (but aggressive) action plan to close the gaps identified in program audits and evaluations. The existing facility should also assess the facility process safety culture and take corrective action for process safety culture improvement, as appropriate.

Broad variations in substance and quality are not uncommon. Within your own company, some businesses, because of their individual needs, may have addressed a number of PSM issues ahead of other parts of the company. In the course of the baseline assessment, you may well identify programs that exceed the minimum requirements of your PSM goals and objectives, as well as those that fall short.

Expect broad variations in style. Many companies (and individual facilities) have both formal and informal management systems. The degree of formality (e.g., documentation, authorization, follow-up) is often a function of the degree of risk associated with the activity the management system is designed to control, as well as the structure or culture of the organization itself.

In the absence of explicit criteria, the seven characteristics listed below may be useful in evaluating management systems. Keep in mind that these characteristics are not absolute requirements; facility management systems may vary significantly and still be capable of achieving the desired results.

Management system characteristics

1. **Policies, programs, and procedures.** Goals and objectives have been set. Formal corporate and facility work plans, policies, guidelines, standards, and procedures are available to clearly define the PSM program scope, outputs, milestones, initiating mechanisms, and alternatives. Management demonstrates commitment to and sponsorship of these programs. Policies, programs, and procedures are periodically reviewed and revised.
2. **Definition of responsibilities.** Facility personnel understand their roles and responsibilities in achieving the desired level of PSM program implementation and process safety performance. Appropriate checks and balances have been established to minimize conflicts of interest. Internal coordination and communication mechanisms exist.

3. **Approvals and authorizations.** Appropriate delegations have been established, and authorities are clearly established for approval of specific routine operations and nonroutine or out-of-specification operations. Variance procedures are defined. Approval levels are commensurate with the importance of the task. Appropriate resources are authorized.
4. **Personnel training and experience.** Facility personnel have sufficient experience, training, and awareness to accomplish the PSM program or activity. Personnel are familiar with applicable regulatory requirements, internal standards and guidelines, and RAGAGEPs. Employees at all levels in the organization are involved in program development.
5. **Protective measures.** Safeguards have been established to prevent or control major problems; administrative controls are in place to cross-check completion of critical operations.
6. **Documentation.** Results of PSM activities are documented, as are compliance/performance results.
7. **Internal verification.** Systems or procedures are in place for reviewing performance against standards and milestones and for reporting departures from established (external or internal) standards.

Process safety performance. In addition to evaluating the PSM performance of an existing facility as described in the acquired facility stage above, the facility personnel should ensure that the corrective action plan is current and that all action items are addressed in a timely manner and tracked to completion. Many facilities include a verification step in the corrective action closure work process, which requires that an independent person review what was recommended and compare that to what was actually done to help ensure that the action was properly addressed before closing the action item in the tracking database.

Similarly, the existing facility should have an established means to ensure that nonconformances identified during MI inspections are expeditiously corrected. Pass/fail tolerances should be set and adhered to. Any decision to operate equipment past its estimated remaining life should be carefully reviewed and endorsed by facility management (including experts or technical authorities).

Finally, analysis of process safety performance should be a significant part of, and integral to, the annual (or periodic) planning that sets goals and objectives, identifies initiatives, and establishes spending for the organization.

Table 2.1 summarizes the techniques that are available for the various process stages.

Table 2.1 Techniques Available to Establish a PSM Program at Various Process Stages

	Facility in design phases	New company or facility	Acquisition of company/facility	Facility expansion	Existing facility
Establish PSM system	X	X	X	X	
Audit PSM system			X	X	X
Define PSM program	X	X	X	X	
Conduct/review audit results		X	X		X
Define lagging indicators	X	X	X	X	
Review lagging indicators		X			X
Define leading indicators	X	X	X	X	
Review leading indicators		X			X
Review performance characteristics			X	X	X

2.4 DOCUMENTING CONCLUSIONS

Once the information needed is gathered, the next challenge is to organize it so that the team can readily identify PSM system gaps and related issues.

One useful tool in understanding the overall status of PSM systems is to rate the management system for each element at each facility using a qualitative rating scale. One approach is to rate PSM system maturity using the following definitions (Ref. 2.5):

Maturity Level 1 programs

- Staff members react in a firefighting mode to the most immediate and pressing need.
- Program effectiveness depends on one or two key people. If the key people leave, a significant part of the program and institutional knowledge goes with them.

- The facility is more or less in compliance with those regulations that are known to it, but there is little assurance that all applicable requirements have been identified, let alone addressed.
- Documentation of compliance is weak.
- Written program documentation is spotty and incomplete, and there is little coordination or correlation among program areas.

Maturity Level 2 programs

- The facility has set some PSM goals.
- Established programs exist.
- The program capabilities are integrated into the organization to a greater degree than in Stage 1. If a key person leaves, the program is likely to recover after a short time.
- The facility can demonstrate compliance with most applicable regulatory requirements, and it has identified those areas where compliance is not yet achieved.
- The facility periodically reviews its compliance status to ensure that the programs that have been implemented are operating as designed.

Maturity Level 3 programs

- The facility sets formal PSM goals and objectives and tracks their progress.
- Staff members have moved beyond managing for compliance and are now actively managing risk.
- Program capabilities are fully integrated into the organization. If a key person leaves, organizational recovery is a function of being understaffed, not underskilled.
- The facility can demonstrate and document compliance with all applicable regulatory requirements.
- Written programs are complete and satisfy the regulatory requirements.
- The facility has a regular self-inspection program utilizing experts from outside the facility or company, in addition to internal personnel.
- New regulatory requirements are anticipated and tracked by the organization, and compliance is achieved according to a regulatory schedule.

The facility ratings can then be combined (see Table 2.2) to indicate overall PSM strengths and weaknesses.

Table 2.2 Example of a PSM Status Summary

PSM Element	Embryonic Systems (Maturity Level 1)	Developing Systems (Maturity Level 2)	Mature Systems (Maturity Level 3)
Management of change		Facility A Facility B	Facility C
Training	Facility A	Facility B Facility C	
Process knowledge and		Facility A Facility C	Facility B
Project review		Facility B	Facility A Facility C
Human factors	Facility A Facility B Facility C		

Regardless of the evaluation method selected, this exercise should provide the team with a clear idea of where the company excels and where it needs work in terms of current PSM implementation and status. These results can be summarized in a concise progress report (see Table 2.3 for an example based on the OSHA PSM model).

Table 2.3 Example of a Progress Report from a PSM Team

Progress Report from the Process Safety Management (PSM) Team

This report summarizes the PSM team's progress in developing PSM systems. Since receiving management approval to proceed, an assessment has been conducted to ascertain the status of PSM in each division. These assessments have found the following:

PSM Element	Status by Facility		
	A	B	C
Employee Participation	+	—	—
Process Safety Information	++	+	—
Process Hazard Analysis	—	—	O
Operating Procedures	++	++	+
Training	++	++	++
Contractors	O	O	O
Pre-startup Safety Review	—	—	O
Mechanical Integrity	+	++	+
Hot Work Permit	+	+	+
Management of Change	+	OO	—
Incident Investigation	+	+	+
Emergency Planning and Response	+	+	+
Compliance Audits	—	+	+
Trade Secrets	+	+	+

++ PSM system in place, documented, and fully operational
\+ Informal system in place
— Incomplete system in place; upgrading needed
O No system in place
OO System exists but is not followed

2.5 REFERENCES

2.1 Center for Chemical Process Safety of the American Institute of Chemical Engineers, *Guidelines for Auditing Process Safety Management Systems, 2nd Edition*, John Wiley & Sons, Inc., Hoboken, New Jersey, 2011.

2.2 Center for Chemical Process Safety of the American Institute of Chemical Engineers, *Guidelines for Process Safety Metrics*, John Wiley & Sons, Inc., Hoboken, New Jersey, 2009.

2.3 ANSI/API Recommended Practice 754, *Process Safety Performance Indicators for the Refining and Petrochemical Industries*, American Petroleum Institute, Washington D.C., April 2010, www.publications.api.org/.

2.4 Covey, Stephen R., *The 7 Habits of Highly Effective People: Powerful Lessons in Personal Change*, Simon & Schuster, Inc., New York, New York, 2013.

2.5 Center for Chemical Process Safety of the American Institute of Chemical Engineers, *Guidelines for Implementing Process Safety Management Systems* (citing Arthur D. Little, Inc.), John Wiley & Sons, Inc., Hoboken, New Jersey, 1994.

3

PREPARING FOR PROCESS SAFETY MANAGEMENT CHANGE

An organization is better prepared for successful PSM change when (1) management commitment is secured, (2) a culture for change is established, and (3) PSM is integrated with HSE and business management systems, wherever possible.

3.1 SECURING MANAGEMENT COMMITMENT

Whether developing a new PSM system or improving an existing one, it will not be effective, efficient, and successful without management commitment. Time invested in gaining management's commitment pays off in tangible support, which not only facilitates, but also expedites, a long-term process such as PSM implementation or improvement. Commitment, as it is used here, refers to explicit, concrete actions, not merely to rhetoric. Having management say, "We are committed to the principles of process safety management" is only the first step. The goal for gaining management commitment is to complete the thought: "We are committed to the principles of process safety management *and we will devote our own time to lead the organization so that these principles are embedded in our operations. We will also provide the necessary staff and financial resources to ensure success.*"

Without personal commitment from the CEO, PSM will not succeed. An initiative that lacks a collective sense of priority and urgency is likely to be carried out piecemeal, despite the best efforts and intentions of its champion(s). Over time, piecemeal implementation is inefficient, since it is likely to take longer and cost more; most importantly, it is very likely to be less effective and successful. PSM is a continuous process, not an event or a series of discrete activities. Without continuity, the implementation or improvement process can easily break down.

Tangible support means not only providing resources, but also ensuring adequate standing relative to other company priorities. PSM needs legitimacy as a business objective to hold its own in situations requiring a trade-off between long-term process improvement and short-term commercial considerations. No matter how deeply committed safety professionals or others supporting or leading PSM may be, this legitimacy can only be conferred by the personal commitment of the CEO and board.

As a practical matter, support from senior management also creates strong incentives at the implementation level. If PSM is known to be a priority for senior management, it is much more likely to attract active participation within the company. By contrast, initiatives that employees see as "flavor of the month" win (and usually deserve) little continuing employee support. In addition, commitment from the top sets standards – and deadlines – for performance. While no one wants to constantly invoke and involve senior management (and it can certainly be counterproductive to do so), specific, articulated expectations from the top greatly improve the chances that individual commitments will be met.

Having established the value of management commitment, the true challenge is to win it. While there is no single, fail-safe formula, there are some identifiable initial steps to consider within the context of your own company.

Section 3.1.1 Defining Senior Leadership Roles

Leaders need to understand the risks posed by their organization's activities and balance major accident risks alongside the other business threats. Even though major accidents occur infrequently, the potential consequences are so high that leaders need to recognize:

- major accidents as credible business risks;
- the integrated nature of many major hazard businesses – including the potential for supply chain disruption; and
- the need for the management of process safety risks to have equal footing with other business processes, including financial governance, markets, investment decisions, etc.

Good PSM needs the active involvement of senior leaders. It is important that they are visible within their organization because of the influence they have on the overall safety and organizational culture.

To maintain the focus on preventing major accidents, leaders also need to recognize the full extent of the impact of these incidents and the potentially devastating consequences for a business, including:

- harm to people, including loss of life and serious injury;
- environmental damage – for example, air, water and land contamination;
- the damage to business efficiency from disruption of production, and the loss of customers or suppliers;
- the potentially huge costs involved – both direct (e.g., asset replacement or repair costs, legal fees, fines) and indirect (e.g., increased insurance premiums; loss of shareholder confidence, resulting in falling share value);
- negative effects on the local economy;

- long-term damage to an organization's reputation due to adverse publicity, legal action, etc.; and
- discontinuation of the company as a viable, ongoing entity in light of the above impacts.

Key self-check questions that senior managers (including the CEO, human resources manager, finance manager, and operations managers) should be able to answer include the following:

- Do you know what the major accident risks are for your organization?
- Do you know what your main vulnerabilities are?
- What are you doing about them?
- How concerned are you about the current level of risk?
- How confident are you that all the safety systems are performing as they should?
- Do you seek out the bad news as well as the good?
- If there is an incident, who do you blame? Others, or yourself?
- Are you doing all you can to prevent a major accident?

3.1.2 Selecting the PSM Champion and the Management Sponsor

The PSM champion is the person responsible for driving the PSM initiative, whether it is the initial implementation of PSM within the organization or an initiative to improve the existing PSM system. The typical PSM champion has a background in safety, engineering, and/or operations because the elements of PSM (whether they are the 12 elements listed in the previous edition of this book [Ref. 3.1], the 20 elements listed in *Guidelines for Risk Based Process Safety* [Ref. 3.2], or the 14 OSHA PSM and EPA RMP elements) involve aspects of each of these disciplines. Familiarity with and understanding of these disciplines improves the credibility and effectiveness of the PSM champion when interfacing with key stakeholders across the organization. The PSM champion can be a corporate, regional, or even facility staff member, depending on the size and structure of the organization. The ideal champion for PSM also knows what makes the company succeed and what broad strategic priorities drive its business. Lastly, the successful champion is skilled at gaining the support of colleagues and building consensus.

Like the PSM champion, selecting the right management sponsor is important for ensuring the success of PSM, whether it is the initial implementation of PSM within the organization or an initiative to improve an existing PSM system. Each company has its own culture, management style, and organizational behavior and structure. In many cases, the functional organization differs from the one illustrated in a formal organizational chart; that's because projects are accomplished by and through a functional organization rather than the formal one. Also, expectations for the sponsor may vary depending on the organization. In some cases, the sponsor will assume a very active, hands-on role, while in others endorsement and oversight will be the sponsor's primary contributions. All of these considerations should be accounted for in selecting the ideal management sponsor for PSM. Depending on the size and structure of the organization, the typical management sponsor for PSM is a corporate or regional leader in operations or safety. Ultimately, the PSM sponsor must be able to ensure that senior management (which owns process safety) is committed to PSM in order to ensure long-term continuity, consistency across the company, and conformance to corporate policy.

3.1.3 Selling the Need for PSM Implementation or Improvement

Having selected the right sponsor and champion, the next task is to sell senior management on the concept of PSM – the need for an effective system within your company, whether the goal is initial development and implementation of PSM or improving an existing system.

Understand your audience

The identified champion should work with the sponsor and, as appropriate, other stakeholders in the organization to understand what senior management – the “audience” – would be looking for. Examine their business priorities, track records with comparable initiatives, and professional backgrounds. The champion should also determine how much detail to provide on the principles of process safety. This may depend on the career paths and professional backgrounds of your company's senior managers. Those who came up through the operations side or have facility-based experience and/or technical training may be more immediately knowledgeable about process safety and its management than others whose backgrounds may be in finance, sales, law, or marketing.

A key point to remember is that senior management's day-to-day priorities are almost certainly different from the champion’s. Their job is to guide the company *as a whole, over the long term.* In determining how best to do this, senior management must consider business operations within the context of a range of factors (economic, social, political) that influence corporate strategy.

Be sure to keep the arguments focused. Everyone has had the frustrating experience of listening to someone who takes forever to get to the point and gets tangled up in irrelevant side discussions. A focused argument targets the listener's interests and agenda, and keeps background and side issues to a minimum. Examples and anecdotes are often useful, but an overreliance on personal experience or "war stories" loses attention. Similarly, an argument that tries to address every conceivable contingency, quirk, and variation will almost certainly fail to convince your audience because of simple overkill.

Frame PSM in terms of the interests of the audience. People pay more attention to information that is relevant to *them* rather than to items of general or academic interest. For example, "PSM provides a cost-effective means of improving our company's safety performance" addresses two senior executive concerns: corporate earnings and exposure to liability. The discussion is framed in terms of the listener's interest; it also focuses on results.

In companies in which senior management has strong operating experience, the concepts behind PSM are more likely to be self-evident. In these cases, limit preliminary discussion of PSM principles to a brief summary. In other companies, senior management's knowledge of operations and process safety may be limited, meaning that the first task is to provide basic information. In cases where there is a mixed audience consisting of those with and without process safety and operations knowledge, the information needs will be different. In these cases, an important task is to seek buy-in up front from the more experienced audience participants to reinforce the suggested messages. In any case, it is obviously important to assess, in advance, the information needs of the people whose endorsement you seek.

Identify PSM benefits

Once an understanding of the audience has been obtained, the next steps are to identify the benefits of implementing a new PSM system or improving an existing one, and then match these benefits to the needs of your audience (e.g., senior management). *While compliance is certainly a requirement in deciding to implement or improve PSM, it is by no means the only benefit.* Rather, compliance is the baseline from which other benefits evolve. An effective PSM system offers benefits over and above simply complying with regulations.

In addition to ensuring employee safety and compliance, the primary benefits of implementing and improving PSM systems include:

- reducing the probability of a fatality or major environmental event,
- avoiding damage to the company's reputation and value (i.e., share price, potential resale price for private companies),
- avoiding nonpayment of bonuses to senior management that would otherwise occur after a major accident resulting in fatalities and significant operational downtime, and
- avoiding senior management distractions from incident investigation and follow-up activities needed after a major accident or fatalities.

Some additional, secondary potential benefits of implementing and improving PSM systems include the following:

- Reduced costs and downtime through:
    - improved maintenance practices and systems that reduce the frequency of equipment failure, improve process reliability, and improve maintenance planning
    - managing process changes to avoid process upsets and downtime

  - o improved operations information, allowing you to track variances in process operating conditions, ensuring fewer rejects and less rework or waste
- Capital cost savings from the systematic review of new projects and identifying safety, reliability, and design enhancements early in the project design phase
- Improved customer satisfaction resulting from enhanced quality
- Increased prestige within industry and among shareholders
- Improved employee recruitment and retention though a clear commitment to safety and consistent management practices
- Improved labor relations by involving union leadership in PSM and consistently communicating with hourly personnel

The exercise of identifying the potential benefits for your company can yield the core of your sales message for developing or improving a PSM system. As can be seen from the list above, PSM has multiple benefits for both specific operations and the company overall. Taking the time to do this up front helps focus the initiative and frame the rationale required to obtain senior management buy-in. In addition, if you (the champion and the sponsor) include others in the exercise, you can build the foundation for future cooperation. Taking a disciplined approach to identifying benefits also helps anticipate questions from, and the concerns of, senior management and others.

A good resource for helping identify the benefits of PSM and for ways to improve PSM systems is the *Process Safety Beacon* produced by CCPS (Ref. 3.3). The monthly one-page *Process Safety Beacon* covers a wide range of process safety issues. Each issue describes a real-life accident and discusses the lessons learned and practical means to prevent a similar accident.

Make your case and support it

Once you have identified and prioritized the PSM benefits for your company, they must be organized into an effective presentation. Presentation formats vary by company, so there is not a single "right" way to do this. Some companies are very formal, requiring written agendas and leave-behind documents, and expecting a very structured, scripted presentation with questions and answers at the end. Others are more free-wheeling, with presentations taking the form of group discussions structured loosely around a topic outline.

It is up to you to determine your company's style and your management's preferences, and to develop a presentation that best meets their needs and expectations. However, there are some fundamental presentation development and delivery techniques that may be useful to you. The following ideas are equally applicable regardless of what form of sponsorship you seek or what organizational structure applies to your company:

1. *Start with an outline* to keep your proposal focused and to ensure that all the key points are covered.
2. *Prepare your proposal for the benefit of key people who may not be present* to ensure their understanding of your proposal.
3. *Use examples* to illustrate why developing or improving a PSM system is important, and note where there is a direct correlation between suggested PSM improvement activities and specific actions resulting from incident investigations.
4. *Prepare an executive summary* for your proposal that includes the course of action, its rationale, and the next steps.
5. *Rehearse* your presentation to ensure that you effectively make the case for PSM.

3.1.4 Selling the Need for Top-level Commitment

All right, says senior management, you've convinced us that PSM is a good idea, and we like your ideas. Now, what do you want from us? Alternatively: All right, says senior management, we're sold; go do it. What do you need us for?

Define the need

Answering these questions requires having a clear idea of what role you want senior management to play and being able to articulate very specific recommendations for their participation. As discussed in previous sections, the role of senior management in implementing PSM may vary from company to company, reflecting differences in style and structure. However, there are four general ways in which top management can be valuably deployed, regardless of individual structure or style:

1. *Lead from the top.* Communicate the process safety mission statement to all staff and contractors. Make process safety the first agenda topic of board meetings. Put on personal protective equipment (PPE) and walk around to discuss process safety issues with front-line staff during every visit to an operational site. Consider other similar activities.
2. *Set corporate goals for PSM.* These may be tangible (e.g., dollar savings, percent reduction in accidental releases) and/or philosophical (e.g., "Our company will be an industry leader in process safety").
3. *Communicate the importance of PSM.* This includes internal as well as external communications and may be part of a broader corporate communications strategy or a freestanding effort.
4. *Provide resources for PSM.* This means recognizing that PSM will require some level of investment and authorizing appropriate allocations of staff and other resources to achieve PSM goals.

In addition, some common denominators probably apply across the board:

- *Senior management's role supports but does not duplicate the efforts of the PSM champion.* The boss should understand that you do not expect anyone to do your job for you. At the same time, senior management should recognize that there are some activities, such as allocating resources or representing the company in a high-level business forum, that you cannot appropriately undertake.
- *Senior managers generally respond better to specific requests than to broad expressions of need.* "First we need a mission statement" is likely to draw a quizzical expression and limited assistance. "The task group has developed this mission statement for your review and signature; we plan to distribute it to all facility locations to kick off the effort" tells the boss exactly what you need and why you need it, and establishes that you expect to carry the ball.
- *Top-level support is often most valuable when it is highly visible.* Perhaps the most useful role senior management can play on behalf of PSM is to endorse it explicitly and visibly, both inside the company and externally. Senior managers' active participation in communications about PSM lends credibility and generates awareness of PSM as a company priority in ways that not even the most dedicated staff team can achieve.

3.2 ESTABLISHING A CULTURE FOR CHANGE

Implementing a new PSM system or improving an existing one can be viewed as an organizational change. The organizational change for PSM can involve (1) creating a new culture (in new companies, those built through acquisitions, or those developing and implementing PSM for the first time), (2) combining cultures (in company mergers or where facility or regional PSM standards are being merged into regional or corporate PSM standards, respectively), or (3) reinforcing or reviving cultures (in older companies, ones that have deviated from their foundational principals, or ones that are improving an existing PSM system).

A culture for change is one in which the organization can accept and adopt changes and prepare employees and contractors working in the company to individually accept and adopt the change(s), empowering them to move out of the current status quo and toward the behavior and belief that represents the desired culture – a culture for change. Ultimately, culture trumps vision, in the sense that the company culture enables employees and contractors to progress on the pathway toward obtaining the vision for PSM. Vision is where we are going as an organization or company; culture determines how and if we get there.

There are three basic steps to changing the culture of an organization:

1. *Become aware of and assess the current culture.* Culture is the shared values and beliefs that drive the behavior of individuals within an organization. It is based on the shared history of these individuals.

Ultimately, it is "the way things are done" in an organization. In order to change the culture of an organization, the first step is to become aware of and assess the strengths of the culture and the challenges the current culture presents to change. The files on the Web accompanying these guidelines includes one example of a safety culture survey.

2. *Envision the new culture.* What does the new culture look like? How is it different from the current/past culture? How can each individual in the organization make a difference? Define a purpose to fuel the vision (e.g., PSM implementation or improvement that can also lead to some or all of the additional benefits listed in Section 3.1.3 above). Align with your leadership team and share the vision with everyone.
3. *Model and implement the culture you want.* Positively and enthusiastically embrace the vision of the new culture. Leaders need to communicate explicitly about the new culture and the underlying behaviors that will best support the new way of doing business. Be prepared for the unexpected and navigate adversity. Remember that while culture changes slowly, it is changing all the time based on each decision made by leadership and the collective actions of the individuals within the organization.

Also, see Chapter 8 of these guidelines for more information and references pertaining to process safety culture.

3.2.1 Key Aspects of Change Management

Some key aspects of change management are:

- *leadership and commitment* from senior management, employees, and contractors;
- the actual *resources committed to executing the change*;
- *communication* of the change to employees and contractors; and
- the *time and duration involved in implementing and sustaining the change.*

Leadership and commitment

The change process for developing or improving a PSM system starts with leadership and commitment from senior management, which is demonstrated by setting goals for PSM, communicating the importance of PSM, and providing resources for PSM, as previously discussed. Through goal-setting, communication, and providing resources for PSM, senior management aligns the values of the organization, and the principles that represent those values, to affect the desired change to implement or improve PSM. Top management support and a leadership style that emphasizes team building and getting employees involved in project development (i.e., allowing employees to have ownership of the project) can lead to greater success. Without this demonstrated leadership and commitment from senior management, the initial development and implementation of a new PSM system or the effort to improve an existing system will not be as successful as possible, and any other potential benefits will not be fully realized.

Leadership and commitment need to also be demonstrated by those employees and contractors responsible for developing, implementing, or improving the PSM system and, ultimately, those affected by the PSM system (i.e., the "change") who are charged with executing the requirements of the PSM system. Thus, it is important to ensure that responsibilities for PSM are assigned to the correct roles within the organization. The PSM system, through the development of supporting policies, procedures, management systems, and practices, can be set up to enable the desired performance of affected employees and contractors by accounting for their current responsibilities and demands (see Chapter 8 of these guidelines for more information). To be successful, the PSM system needs to work with and complement other business requirements and management systems related to operations, maintenance, engineering, procurement, and product quality (see Chapter 7 of these guidelines for more information). When the system is set up correctly, individuals quickly realize that the change (1) is beneficial to them, (2) can help them individually in their jobs, and (3) can help the company ultimately remain competitive in the marketplace. As with senior management, the success of PSM and the realization of its other potential benefits depend on the leadership and commitment of employees and contractors.

Resources committed to executing the change

It almost goes without saying that the success of change depends on the resources committed to the change by senior management. Typically, we think of resources as the budget for the project and the number of people on the project team. The skills, knowledge, and experience of the project team also influence the time needed to implement the change and the long-term success and sustainability of the change.

Communication

Communication is important to helping senior management demonstrate commitment and gaining the commitment of all employees and contractors. An entire communication program for implementation of the project, from beginning to end, and for ongoing support of the program is needed to involve and enable people within the organization. Communication of the change should begin as early as practical and be as open and comprehensive as possible to gain and keep the commitment of employees and contractors.

Sites communicate in different ways, but a good PSM system in an operating facility would typically involve communication activities such as quarterly presentations to all on site (via town-hall meetings, video presentations, etc.) that cover the most recent process safety incidents, KPIs, and the status of projects currently underway that address process safety issues. These presentations need to be delivered by the most senior person available and not delegated to the PSM or HSE manager.

Time and duration involved in implementing and sustaining the change

All changes involve a time element or duration. How long will it take to implement the change? Is the change permanent or temporary? To get an idea of the implementation time, you need to understand where the organization is at the moment, where it wants to be and by when, and what measures need to be taken to get there. Then develop a plan to achieve the desired end state that includes achievable and measureable milestones and timing targets.

3.2.2 An Example Change Management Approach

One example of a change management approach is John P. Kotter's "8-Step Process for Leading Change" (Ref. 3.4):

1. *Establishing a sense of urgency* to help others see the need for change and they will be convinced of the importance of acting immediately.
2. *Creating the guiding coalition* by assembling a group with enough power to lead the change effort and encourage the group to work as a team.
3. *Developing a change vision* to help direct the change effort and develop strategies for achieving that vision.
4. *Communicating the vision for buy-in* to make sure as many people as possible understand and accept the vision and the strategy.
5. *Empowering broad-based action* to remove obstacles to change, change systems or structures that seriously undermine the vision, and encourage risk-taking and nontraditional ideas, activities, and actions.
6. *Generating short-term wins* by planning for achievements that can easily be made visible, following through with those achievements, and recognizing and rewarding employees who were involved.
7. *Never letting up* by using the increased credibility to change systems, structures, and policies that don't fit the vision, also hire, promote, and develop employees who can implement the vision, and finally reinvigorate the process with new projects, themes, and change agents.
8. *Incorporating changes into the culture* by articulating the connections between the new behaviors and organizational success, and developing the means to ensure leadership development and succession.

Again, this is one example of a change management approach. There are many books and articles available on change management and establishing a culture of change that can be found through a simple Web search (Refs. 3.5 through 3.9).

3.3 REFERENCES

3.1 Center for Chemical Process Safety of the American Institute of Chemical Engineers, *Guidelines for Implementing Process Safety Management Systems,* John Wiley & Sons, Inc., Hoboken, New Jersey, 1994.

3.2 Center for Chemical Process Safety of the American Institute of Chemical Engineers, *Guidelines for Risk Based Process Safety*, John Wiley & Sons, Inc., Hoboken, New Jersey, 2007.

3.3 Center for Chemical Process Safety of the American Institute of Chemical Engineers, *Process Safety Beacon*, New York, New York, www.aiche.org/ccps/resources/process-safety-beacon.

3.4 Kotter, John P., *Leading Change*, Harvard Business Review Press, Boston, Massachusetts, 2012.

3.5 Krause, Thomas R., *Leading with Safety*, John Wiley & Sons, Inc., Hoboken, New Jersey, 2005.

3.6 Hyatt, Michael, "How Do You Change Organizational Culture?," February 22, 2012, www.michaelhyatt.com.

3.7 Jones, John, DeAnne Aguirre, and Matthew Calderone, "10 Principles of Change Management," April 15, 2014, www.strategy-business.com.

3.8 Organisation for Economic Co-operation and Development, *Corporate Governance for Process Safety: Guidance for Senior Leaders in High Hazard Industries*, OECD Environment, Health and Safety, Chemical Accidents Programme, Danvers, Massachusetts, June 2012, www.oecd.org/chemicalsafety/corporategovernanceforprocesssafety.htm.

3.9 Sirkin, Harold L., Perry Keenan, and Alan Jackson, "The Hard Side of Change Management," *Harvard Business Review*, Boston, Massachusetts, October 2005, www.hbr.org/2005/10/the-hard-side-of-change-management#.

4

IMPLEMENTING A NEW PSM SYSTEM

Implementing a new PSM system includes the following steps:

1. Develop the design specification for the PSM system (by reviewing existing frameworks to determine the one preferred going forward). This includes performing a gap assessment between the PSM system to be implemented and existing processes and procedures.
2. Create element and system workflows (as appropriate).
3. Estimate the workloads and resources needed to implement the new elements and system.
4. Develop written programs and procedures for the elements and system.
5. Roll out the elements and system at a single site to act as a pilot program before rolling them out to the entire company.
6. Monitor implementation and initial performance, and modify the elements and/or system to make them work for the pilot site. Once the PSM system is rolled out to the entire company, monitor its progress every six months and share the results with management.

4.1 DEVELOP THE DESIGN SPECIFICATION FOR THE PSM SYSTEM

The first step in implementing any new management system should be developing the design specification for the system. This step is often overlooked in designing management systems, but it is just as important as in designing a chemical process. This step should consider (1) the PSM system design parameters that are important, (2) how the system should interface with existing product and capital execution processes, and (3) the desired element/system design characteristics.

4.1.1 Select the PSM System Structure

PSM system design should consider:

- the framework (or model) that will be used, including all the associated PSM elements;

- the foundation and starting point for each element (i.e., Are there existing systems that provide a foundation, or is this a brand new element?); and
- the level of detail that will be provided for the design of each element.

Select the PSM framework

As a first step, you should decide which framework (or model) to use. A framework defines the elements and components of your program, and it must be established before you can (1) evaluate what is already in place and (2) plan for filling the gaps.

During the past 20-plus years, several descriptions have emerged of what elements should be included in a PSM system. You should consider these and select one (or define your own based on these). In selecting among the available alternatives, there are several factors to consider:

- **Preexisting company frameworks.** If you already have your own framework for PSM, it is advantageous to retain it if it has proven effective. The existing system will be readily understood and accepted within the organization. Before choosing a new framework, you should compare the contents of your PSM framework against the alternative models, ensuring that your system is complete. Where necessary, you can modify your existing PSM framework to include additional elements (see Chapter 5 of these guidelines for more information).

  Note: An easy way to compare frameworks is to begin with a list of the components of (for example) the CCPS RBPS framework; then, next to each RBPS component, list the corresponding component of your framework.
- **Commitments to industry programs.** If your company is committed to an industry initiative, such as API's Recommended Practice (RP) 750 or ACC's Responsible Care® program, you will want to ensure that you are consistent with the framework used by that program.
- **Flexibility among divisions/operations.** If your company has divisions or operations involved in different types of products, you will likely want to use a framework that is broadly adaptable among these different operations.
- **Management system design.** You should use a framework that helps define what a management system is, so that you will be better able to ensure that the systems you design are comprehensive. In other words, you should be looking for more than a list of the areas of concern (e.g., management of change). Your framework should provide guidance on what the management system for each element will address.

- **Peer company programs.** You may wish to consider what peers and/or industry leaders are doing. Benchmarking is an approach to examining the ways in which particular business processes are performed by other companies. At this stage of your PSM implementation effort, you may wish to gather some information from other companies. This will help you gain perspective on what others have done, and will also provide "ammunition" for responding to any skeptics within your company who may believe that PSM efforts are consuming or will consume too much effort.

  Note: Many companies are willing to share information about their PSM programs with other firms. CCPS and industry associations provide good opportunities for networking to achieve this. Attending process safety conferences (e.g., the annual Global Conference on Process Safety) is a good way to meet people who share your interest in PSM.

- **Compliance.** In many countries some aspects of PSM are governed by regulations, and these may need to be addressed in your framework. If you are not already in compliance, this will also probably influence your priorities.

With these criteria in mind, you should evaluate the alternative PSM models. A brief review of some of the major alternatives follows.

CCPS PSM models

The first CCPS model for a PSM management system was provided in *Chemical Process Safety Management – A Challenge to Commitment* (Ref. 4.1) and later explained in "Guidelines for Technical Management of Chemical Process Safety" (Ref. 4.2). This model described a PSM management system in terms of 12 elements and 68 components. Table 4.1 summarizes these 12 elements. The model was designed to be applicable throughout the process industries, and it is also applicable beyond process safety to other areas of safety, health, and environmental protection.

The second CCPS model for a PSM management system is provided in *Guidelines for Risk Based Process Safety* (Ref. 4.3). The RBPS model, which is intended to be the framework for the next generation of process safety management:

- builds upon the original CCPS model;
- integrates industry lessons learned over the intervening years;
- applies the management system principles of Plan-Do-Check-Adjust;
- organizes them in a way that will be useful to all organizations – even those with relatively lower hazard activities – throughout the life cycle of a process or operation;

- is intended to be integrated with other elements of company management systems so that it is totally consistent with manufacturing operations, SHE controls, security, and related technical and business areas; and
- may create a new performance-based expectation for process safety.

The RBPS model encompasses 20 RBPS elements, organized under the 4 accident prevention pillars shown in Table 4.2. These 20 elements expand upon the original CCPS model PSM elements to reflect many years of PSM implementation experience, best practices from a variety of industries, and worldwide regulatory requirements. For each of the 20 elements, the RBPS book element chapters include (1) an overview of the element, (2) key principles and essential features, (3) work activities and implementation options, (4) performance and efficiency improvement examples, (5) possible metrics, and (6) management review topics.

The RBPS approach recognizes that all hazards and risks are not equal, and therefore focuses more resources on higher hazards and risks. The main emphasis of the RBPS approach is to put just enough energy into each activity to meet the anticipated needs for that activity. It emphasizes (1) basing process safety improvement efforts on RBPS criteria, (2) measuring performance and efficiency so organizations can apply finite resources in a prioritized manner to a large number of competing process safety needs, and (3) implementing four accident prevention pillars (i.e., commit to process safety, understand hazards and risks, manage risk, and learn from experience) at a risk-appropriate level of rigor.

Section 5.3 of this book discusses the "new" elements in the RBPS model, along with a general approach and specific steps for implementing each of them.

Table 4.1 1995 CCPS PSM System Elements

- Accountability: Objectives and Goals
- Process Knowledge and Documentation
- Capital Project Review and Design Procedures (for new or existing facilities, expansions, and acquisitions)
- Process Risk Management
- Management of Change
- Process and Equipment Integrity
- Human Factors
- Training and Performance
- Incident Investigation
- Standards, Codes, and Laws
- Audits and Corrective Actions
- Enhancement of Process Safety Knowledge

Table 4.2 CCPS's Risk Based Process Safety Elements

Commit to Process Safety
• Process Safety Culture • Compliance with Standards • Process Safety Competency • Workforce Involvement • Stakeholder Outreach
Understand Hazards and Risk
• Process Knowledge Management • Asset Integrity and Reliability • Contractor Management • Training and Performance Assurance • Management of Change • Operational Readiness • Conduct of Operations • Emergency Management • Hazard Identification and Risk Analysis
Manage Risk
• Operating Procedures • Safe Work Practices
Learn from Experience
• Incident Investigation • Measurement and Metrics Auditing • Management Review and Continuous Improvement

These CCPS models are unique in their descriptions of what constitutes a management system, and in their effort to address the planning, organizing, implementing, and control aspects of PSM systems.

Other models

A variety of other descriptions of PSM systems have been published since 1988. Among these are frameworks published by ACC, API, OSHA, EPA, and the European Union (EU). The components of these systems are listed in Tables 4.4 through 4.8, and information about obtaining copies of them is provided in Appendix I.

The UK's Energy Institute also recently developed a "High Level Framework for Process Safety Management" (Ref. 4.4). It was developed by the energy sector with participants being process safety professionals mainly from the offshore and onshore oil and gas and the power sectors. It was intended to capture the practices and experience of the participants implementing other PSM systems, including the RBPS within an existing management system, and incorporate their learnings into the framework.

All of the above approaches arrive at a very similar endpoint. The components of these different frameworks have been combined and realigned to provide the common overview as shown in Table 4.3.

Appendix I of this book also provides information on other PSM system models that are not specifically included in this section.

Comparison of these models shows that they are all very similar. There are differences in terminology and emphasis, but the fundamental concepts of PSM are consistent. For example:

- ACC's Process Safety Code of Management Practices is oriented toward the chemical industry and describes the elements of a PSM program as part of its Responsible Care® program.
- API's RP 750 describes PSM elements recommended for operations in the oil and gas industry.
- EPA's RMP rule 40 CFR Part 68 and OSHA's PSM regulation 29 CFR 1910.119 provide regulatory frameworks for process safety management. These regulations apply to facilities handling specified chemicals above defined threshold quantities.

  Note: Although EPA's prevention program elements are virtually identical to OSHA's PSM elements, the EPA RMP rule has additional requirements (especially regarding hazard assessments and risk management plans) that are above and beyond the OSHA requirements.

  Note: The OSHA and EPA models (and requirements) have not changed significantly since their promulgation in the early to mid-1990s. However, as a result of Executive Order 13650 issued in August 2013, both agencies issued Requests for Information in 2014, asking for input on a number of possible additions and revisions to their regulations. Therefore, readers in the U.S. should be aware of these possible changes to 29 CFR 1910.119 and/or 40 CFR Part 68 (see www.osha.gov and www.epa.gov for the latest information and any updates).

Table 4.3 Overview of Different PSM Frameworks

CCPS RBPS Elements (2007)	CCPS PSM System Elements (1995)	ACC Process Safety Code Elements	API PSM Elements	OSHA PSM Elements	EPA Risk Management Program Components	EI High Level Framework for PSM Elements (2010)
Commit to Process Safety						**Process Safety Leadership**
		Comprehensive Process Safety Management System				
Process Safety Culture	Accountability: Objectives and Goals	Leadership and Culture			Management	Leadership, Commitment and Responsibility
Compliance with Standards	Standards, Codes, and Laws	Accountability				Identification and Compliance with Legislation and Industry Standards
						Standards and Practices
Process Safety Competency	Enhancement of Process Safety Knowledge	Knowledge, Expertise and Training	Training	Training	Training	Employee Selection, Placement, Competency and Health Assurance
Workforce Involvement				Employee Participation	Employee Participation	Workplace Involvement
Stakeholder Outreach						Communication with stakeholders

Table 4.3 ***Continued***

CCPS RBPS Elements (2007)	CCPS PSM System Elements (1995)	ACC Process Safety Code Elements	API PSM Elements	OSHA PSM Elements	EPA Risk Management Program Components	EI High Level Framework for PSM Elements (2010)
Understand Hazards and Risk						**Risk Identification and Assessment**
Process Knowledge Management	Process Knowledge and Documentation	Information Sharing	Process Safety Information	Process Safety Information	Process Safety Information	Documentation, Knowledge and Records Management
Asset Integrity and Reliability	Process and Equipment Integrity		Assurance of the Quality and Integrity of Critical Equipment	Mechanical Integrity	Mechanical Integrity	Inspection and Maintenance
						Management of Safety Critical Devices
Contractor Management				Contractors	Contractors	Contractor and Supplier Selection and Management
Training and Performance Assurance	Training and Performance					
Management of Change	Management of Change		Management of Change	Management of Change	Management of Change	Management of Change and Project Management
Operational Readiness			Pre-startup Safety Review	Pre-startup Safety Review	Pre-startup Review	Operational Readiness and Process Startup

Table 4.3 *Continued*

CCPS RBPS Elements (2007)	CCPS PSM System Elements (1995)	ACC Process Safety Code Elements	API PSM Elements	OSHA PSM Elements	EPA Risk Management Program Components	EI High Level Framework for PSM Elements (2010)
Conduct of Operations			Safe Work Practices	Operating Procedures	Operating Procedures	Process and Operational Status
Emergency Management			Emergency Response and Control	Emergency Planning and Response	Emergency Response Program	Emergency Preparedness
Hazard Identification and Risk Analysis	Capital Project Review and Design Procedures (for new or existing facilities, expansions, and acquisitions)	Understanding and Prioritization of Process Safety Risks	Process Hazard Analysis	Process Hazard Analysis	Process Hazard Analysis	Hazard Identification and Risk Assessment
					Risk Management Plan	
Manage Risk						**Risk Management**
Operating Procedures	Human Factors		Operating Procedures			Operating Procedures and Manuals
						Management of Operational Interfaces
Safe Work Practices				Safe Work Practices Hot Work Permit	Safe Work Practices Hot Work Permits	Work Control, Permit to Work and Task Risk Management

Table 4.3 ***Continued***

CCPS RBPS Elements (2007)	CCPS PSM System Elements (1995)	ACC Process Safety Code Elements	API PSM Elements	OSHA PSM Elements	EPA Risk Management Program Components	EI High Level Framework for PSM Elements (2010)
Learn from Experience						**Review and Improvement**
Incident Investigation	Incident Investigation		Investigation of Process Related Incidents	Incident Investigation	Incident Investigation	Incident Investigation
Measurement and Metrics						
Auditing	Audits and Corrective Actions		Audit of Process Hazards Management Systems	Compliance Audits	Compliance Audits	Audit, Assurance, Management Review and Intervention
Management Review and Continuous Improvement	Process Risk Management	Monitoring and Improving Performance				
Other						
				Trade Secrets		

Table 4.4 American Chemistry Council Process Safety Code Elements

1.	**Leadership and culture.** Senior leadership commitment to creating and valuing a process safety culture. Each company's leadership will demonstrate a visible and ongoing commitment to overseeing and improving process safety performance.
2.	**Accountability.** Establishment of process safety accountability within the company. Process safety is integral to business processes and stakeholder expectations.
3.	**Knowledge, expertise and training.** Processes to provide that companies and their employees have the required knowledge, expertise, tools and training to manage process risks of their operations.
4.	**Understanding and prioritization of process safety risks.** Processes to systematically understand process safety risks throughout the organization, prioritize actions and allocate resources.
5.	**Comprehensive process safety management system.** Development and documentation of a comprehensive process safety management system to manage process risk and drive continuous improvement.
6.	**Information sharing.** Systems to actively share relevant process safety knowledge and lessons learned across the organization, including methods for making information available to relevant stakeholders.
7.	**Monitoring and improving performance.** A system to monitor, report, review and improve process safety performance

Table 4.5 American Petroleum Institute Process Safety Management Elements

- Process Safety Information
- Process Hazard Analysis
- Management of Change
- Operating Procedures
- Safe Work Practices
- Training
- Assurance of the Quality and Integrity of Critical Equipment
- Pre-Startup Safety Review
- Emergency Response and Control
- Investigation of Process Related Incidents
- Audit of Process Hazards Management Systems

Table 4.6 Occupational Safety and Health Administration (OSHA) Process Safety Management Elements

- Employee Participation
- Process Safety Information
- Process Hazard Analysis
- Operating Procedures
- Training
- Contractors
- Pre-Startup Safety Review
- Mechanical Integrity
- Hot Work Permit
- Management of Change
- Incident Investigation
- Emergency Planning and Response
- Compliance Audits
- Trade Secrets

Table 4.7 Environmental Protection Agency (EPA) Risk Management Program Components

- Management
- Hazard Assessment
- Prevention Program (Program 3)
  - Process Safety Information
  - Process Hazard Analysis
  - Operating Procedures
  - Training
  - Mechanical Integrity
  - Management of Change
  - Pre-Startup Review
  - Compliance Audits
  - Incident Investigation
  - Employee Participation
  - Hot Work Permits
  - Contractors
- Emergency Response Program
- Risk Management Plan

Table 4.8 EU Seveso Directive Safety Management Elements to Prevent Major Accidents

(i)	organization and personnel
(ii)	identification and evaluation of major hazards
(iii)	operational control
(iv)	management of change
(v)	planning for emergencies
(vi)	monitoring performance
(vii)	audit and review

Comparison of these models shows that they are all very similar. There are differences in terminology and emphasis, but the fundamental concepts of PSM are consistent. For example:

- ACC's Process Safety Code of Management Practices is oriented toward the chemical industry and describes the elements of a PSM program as part of its Responsible Care® program.
- API's RP 750 describes PSM elements recommended for operations in the oil and gas industry.
- EPA's RMP rule 40 CFR Part 68 and OSHA's PSM regulation 29 CFR 1910.119 provide regulatory frameworks for process safety management. These regulations apply to facilities handling specified chemicals above defined threshold quantities.

  Note: Although EPA's prevention program elements are virtually identical to OSHA's PSM elements, the EPA RMP rule has additional requirements (especially regarding hazard assessments and risk management plans) that are above and beyond the OSHA requirements.

  Note: The OSHA and EPA models (and requirements) have not changed significantly since their promulgation in the early to mid-1990s. However, as a result of Executive Order 13650 issued in August 2013, both agencies issued Requests for Information in 2014, asking for input on a number of possible additions and revisions to their regulations. Therefore, readers in the U.S. should be aware of these possible changes to 29 CFR 1910.119 and/or 40 CFR Part 68 (see www.osha.gov and www.epa.gov for the latest information and any updates).

Foundation and starting point for each element

The foundation and starting point for each PSM element will fundamentally depend on whether there are existing systems, functions, or activities that provide a foundation for some of the design characteristics, or whether this is a brand-new element. Where there are existing systems or parts of systems in place with proven effectiveness, utilizing them should help the implementation effort be more successful. In either case, the existing systems (or lack thereof) should be clearly identified and regarded as key considerations when determining (1) the degree of rigor required for each PSM element, (2) the effort that will be required to create the desired systems, and (3) the activities that should be planned to support the implementation of each element.

Level of detail required for each element

The design specification for each PSM element will also depend on the level of detail required for each element. As pointed out in the RBPS book, the appropriate level of detail for effective process safety improvement depends on (1) (primarily) the hazards and level of risk associated with facilities or operations, (2) the level of demand for the activity, (3) the resources needed, and (4) the embedded safety culture.

For this reason, each of the possible work activities listed in Section 4.1.3 and the multitude of examples in the RBPS book provide several "levels" for each activity. The level chosen should depend on the considerations listed in the previous paragraph and the presence of any existing systems (or lack thereof).

4.1.2 Identify Interfaces with Existing New Product and Capital Project Execution Processes

Before getting deeply involved in the design of a new PSM system, it is important to determine what interfaces there currently are for existing new product and capital project execution processes, and what they should look like in the new system. Because these interfaces are often managed, to some extent, by groups that are outside the normal chain of command, they can sometimes be overlooked and/or not given adequate attention, thereby leaving gaps in the PSM system.

New products

New product interfaces with the PSM system generally arise from two sources:

1. New blends or formulations within an existing production operation. This is particularly common in batch operations, blending operations, and production of polymers involving comonomers.
2. Brand-new products, usually requiring new equipment or a new equipment configuration to produce.

In either case, there are a number of potential interfaces with the PSM system that warrant review. Here are a few of the significant questions to ask:

- Has adequate PSI (e.g., hazards of new chemicals involved, chemical reactivity hazards, impact on safe upper and lower limits) been obtained and evaluated?

Note: Understanding potential reactivity hazards is particularly important when handling new products and/or new ingredients in a blend. See the CCPS book entitled Essential Practices for Managing Chemical Reactivity Hazards (Ref. 4.5) *for more information.*

- Is a PHA required? If so, has it been performed and the resulting issues addressed? This is clearly warranted for new products, but what are the criteria for a new blend or formulation?
- Are new operating procedures and/or new operating limits (and training on them) needed?
- How do new products integrate with the management of change (MOC) system? An MOC is clearly warranted for a new product, but what are the criteria for a new blend or formulation?
- Can the new chemicals, intermediates, or products impact the mechanical integrity of the process? Has an MOC been performed and the results evaluated? Have the resulting issues been addressed?

Capital projects

Capital projects are essentially "large" changes from a PSM perspective. They must not be excluded from MOC evaluation, and proper process knowledge must be obtained and maintained. However, the MOC system used to manage most day-to-day changes may be overly cumbersome and difficult to apply to capital projects. In addition, such projects can result in organizational changes that can have large impacts on the PSM system if not properly managed. Therefore, during the design phase of any new PSM system, it is appropriate to consider how the principles of MOC can be applied to capital projects, particularly large ones, in an effective and efficient way. See Chapters 8 and 15 of the CCPS RBPS book, and the CCPS book *Guidelines for Management of Change for Process Safety*, for guidance on activities to consider in order to achieve these objectives.

4.1.3 Define the Element/System Design Characteristics

Considering the factors previously discussed, the design characteristics for the overall systems and for each PSM element should be defined. The RBPS book provides information on possible key principles and essential features, as well as possible work activities, for each RBPS element. Following are examples of work activities for two RBPS elements (Process Knowledge Management and Operating Procedures).

Note: The activities under each item reflect an increasingly rigorous PSM system design.

Process Knowledge Management (see pages 187 through 195 of the RBPS book for the complete list of activities)

Define the scope

2. The written policy should specify the scope of the knowledge element, including the various types of information and documentation that should be created/compiled for each unit at the facility.

   a. Process knowledge is compiled and maintained on an ad hoc basis.
   b. The written policy describes what process knowledge is required for designated areas at the facility.
   c. The written policy clearly describes what process knowledge is required for each unit or process area.
   d. The written policy clearly describes what process knowledge is required for each unit or process area, along with standards regarding how the information will be compiled and stored.
   e. In addition to item (d), the written policy includes checklists or forms that indicate the type of information required for common types of equipment. These forms are tailored based on risk; for example, the requirements for a compressor that is used to supply hydrocarbons to high-pressure process equipment are much more rigorous than for compressors that are used to supply utility air.

Thoroughly document chemical reactivity and incompatibility hazards

5. Include in the written policy governing the *knowledge* element a specific standard for documentation of chemical reactive hazards.

   a. Documentation of chemical reactivity hazards is generally limited to SDSs.
   b. A written policy requires that units maintain SDSs for all chemicals present at the unit, and that data be recorded using a specific form/matrix to summarize the hazards of mixing for chemicals that are normally present at the unit (including utility streams, water, air, and any common contaminants).
   c. A written policy clearly describes what hazards must be addressed, and references tools that help users evaluate special hazards such as self-reactivity, potential for a runaway reaction, shock sensitivity, potential for spontaneous combustion or a dust cloud explosion, alternate chemical reactions that present a special hazard, and other hazards that are related to physical attributes such as particle size.
   d. In addition to item (c), the written policy requires that chemical reactivity hazards be evaluated using appropriate laboratory methods.

Protect against inadvertent change

16. Assign persons with proper knowledge and experience to review and approve corrections or changes to process knowledge.

a. No review or approval is needed for a process engineer or designer to update the process knowledge. When certain process knowledge, such as P&IDs, is field verified, conditions in the field are generally believed to be correct. Thus, the process knowledge is updated to reflect the current field configuration with minimal review.

b. Differences between process knowledge and equipment in the field are generally brought to the attention of a designated process engineer who determines which is correct, and on that basis appropriate changes are made.

c. Differences between process knowledge and equipment in the field are generally brought to the attention of one or more senior technical persons with recognized expertise in the affected area for resolution.

d. Differences between process knowledge and equipment in the field are brought to the attention of one or more senior technical persons who determine which is correct based on an understanding of the designer's intent. A pattern of significant discrepancies is investigated as a chronic near-miss incident.

Operating Procedures (see pages 261 through 273 of the RBPS book for the complete list of activities)

Control procedure format and content

3. Include in the written policy or description of the management system procedure governing the *procedures* element a list or description of acceptable formats/structure for all operating procedures.

   a. A general written policy exists; however, procedures in different areas follow different formats.

   b. A written policy includes sample formats for operating procedures.

   c. A written policy specifies acceptable formats for various types of operating procedures.

4. Provide guidance on content, including what should not be included in operating procedures. Also include guidance on what information should be included in related documents, such as training or process technology manuals.

   a. A written policy provides general guidance on content.

   b. A written policy specifies what content should be included in operating procedures.

   c. In addition to item (b), the written policy includes sample content for different types of operating procedures with detailed examples.

Address safe operating limits and consequences of deviation from safe operating limits

17. Establish safe operating limits for each process parameter where deviation from the limit is credible and could lead to an unsafe condition. Also, for each safe operating limit, state the potential consequence of exceeding the limit and the steps to avoid deviation or return the process to a safe condition if an excursion outside of the safe operating limits does occur.

    a. A fairly narrow range of limits has been established; operation within these narrow limits ensures safe operation and helps ensure that yield and product quality targets are met. However, the operating procedures do not directly address safe operating limits, consequences of deviation, and steps to take to prevent or mitigate the consequences of exceeding safe operating limits.

    b. Safe operating limits can be found in (or sometimes inferred from) the operating procedures, but they are not always addressed in a clear manner.

    c. Safe operating limits, along with (1) the potential consequences associated with exceeding the limits and (2) the steps operators should take to prevent or mitigate the consequences of exceeding the limits, are clearly described in the operating procedures. These limits are maintained in a special section of the procedures or are otherwise very easy to locate because they are consistently presented in all procedures.

Supplement procedures with checklists

20. Include a checklist in the procedure whenever the sequence of operations is important, or when certain steps must be complete prior to moving to the next phase of operation.

    a. Many checklists are developed, but they are normally not supplemented by procedures.

    b. Procedures and checklists are developed in parallel systems, and they are cross checked as part of the periodic procedure review.

    c. Procedures and checklists are co-developed in a manner that the user can access the checklist, the procedure, or both, as needed.

In addition, the RBPS book and the CCPS books published on many individual PSM or RBPS elements include a wealth of forms and tools that may warrant inclusion in your PSM system. For example, see Table 4.9 (shown as Figure 9.3 in the RBPS book) and Table 4.10 (shown as Table 10.1 in the RBPS book).

Table 4.9 Example Risk Matrix

Risk	Serious danger in immediate area	Serious danger inside battery limits	Serious danger site wide	Serious danger offsite
More than once per year	Action required unless risk ALARP	Action required at first opportunity	Immediate action required	Immediate action required
Once every few years	Action required unless risk ALARP	Action required unless risk ALARP	Action required at first opportunity	Immediate action required
Once in the facility's lifetime	No action required	Action required unless risk ALARP	Action required unless risk ALARP	Action required at first opportunity
Not expected in the facility's lifetime	No action required	No action required	Action required unless risk ALARP	Action required unless risk ALARP

Also, the CCPS books published on many individual PSM or RBPS elements (e.g., PHA, MI, MOC, incident investigation, etc.) contain similar information that can be used directly or adapted to suit site or company needs. Go to www.aiche.org/ccps/resources/publications to browse the available books or search for one on a specific topic or element.

All of these and any other available similar resources should be used to develop the element/system design characteristics for your site or company PSM system.

4.2 CREATE ELEMENT AND SYSTEM WORKFLOWS

For many PSM elements, the steps involved in implementing element activities, and the interactions between the steps and the workgroups involved in the implementation activities, are quite complex. For this reason, developing workflows for such elements is very beneficial to ensure that (1) no steps are overlooked or inadequately considered, (2) all stakeholders and workgroups are appropriately involved in the activity, (3) the activity is performed efficiently and effectively, and (4) the potential for significant problems is minimized during initial implementation of the new PSM element.

Table 4.10 Procedure Formats

Type	*Description*
Narrative	Long narrative paragraphs that provide a detailed account of how a task is performed; paragraphs may not be numbered. Widely used, but the narrative format can be very confusing and difficult to follow; this format should be avoided.
Paragraph	Short, numbered paragraphs, typically with a mixture of commands and passive descriptions. Widely used, and better than the narrative format, but more wordy and generally less useful than the outline format.
Outline	Phrases, sentences, and short paragraphs organized using indentation, varied numbering, and logical grouping of information. Often used for sequential or batch operations.
Playscript	Steps grouped according to who performs them or by logical subtasks. Often used for coordinating activities between two operators or operating units, particularly if the order of steps is important.
T-Bar	Two-column format with basic actions in left column and details, notes, and so forth in right column. Can make the procedures longer, but reduces the “noise” typically contained in action steps and helps highlight special details. Particularly useful for combining a step-by-step procedure with a job safety analysis in a manner that minimizes clutter and the potential for confusion.
Multi-Column	A tabular format with multiple compartments of information. Often used for troubleshooting guides or maps that tie other documents/ procedures together.
Flowchart	A graphical format that is structured with boxes, diamonds, and arrowheads and contains brief action and conditional statements. Often used for troubleshooting guides or transactional procedures, particularly to display decision steps (this format has been demonstrated to be superior for nuclear facility emergency response actions).
Checklist	Brief step descriptions providing basic actions only, typically with spaces for check marks or initials/signatures. Most often used for simple, repetitive operations, such as hazardous material unloading. Particularly useful if the steps are critical to safe operation, as in critical nonroutine operating tasks such as shutdown prior to a turnaround and restart after a turnaround, or if a record of successful operation is desired (e.g., a completed checklist.

In addition, many interactions exist between PSM elements (and their workflows) and other business systems. For example, PSI is a foundation or starting point for many other elements (e.g., PHA, operating procedures, MOC, MI). In these cases, developing system workflows or showing the connections between element workflows can help ensure that they are properly integrated and complement each other.

4.2.1 Developing a Workflow Diagram

Wikipedia defines workflow this way: "A workflow consists of a sequence of connected steps where each step follows without delay or gap and ends just before the subsequent step may begin. It is a depiction of a sequence of operations, declared as work of a person or group, an organization of staff, or one or more simple or complex mechanisms. Workflow may be seen as any abstraction of real work. For control purposes, workflow may be a view of real work in a chosen aspect, thus serving as a virtual representation of actual work. The flow being described may refer to a document or product that is being transferred from one step to another."

As introduced in the previous discussion of their benefits, the keys to developing good workflow diagrams are as follows:

- **Involving all stakeholders and workgroups.** First, think about (1) who is or should be either involved in this activity or aware of it, (2) who is accountable for the activity/results, and (2) who the subject matter experts are. Invite personnel representing all these aspects to help with workflow development.
- **Identifying all the steps that should be involved.** This can begin with group brainstorming, or it can be "seeded" by preparing in advance some steps to discuss or presenting a previous or "generic" diagram for review (see Section 4.2.2).
- **Developing the logical/desired flow between steps.** Next, the group should develop the logical/desired flow between the steps, including considerations such as "recycle loops" based on some results and "handoffs" to/from other PSM elements (and other business systems). Using "sticky" notes and positioning (and repositioning) them on a wall, board, or paper can be a useful technique in quickly considering and resolving possible changes in the workflow.
- **Identifying and understanding inputs/outputs and goals and objectives associated with each step.** It is also a good practice to identify available data (e.g., system reports) that can be used to measure system performance and health (i.e., real-time monitoring; see Section 4.4.8) during workflow development.

- **Critiquing the workflow.** Finally, and perhaps during a subsequent meeting to allow "soak" time and review of the draft workflow, the group should critique the workflow to (1) identify potential problems or issues and (2) develop ideas/recommendations to ensure efficient and effective performance of the activity.

  Considerations should include resources/tools required for each step and for handoffs, potential rough transitions or "dropping" of activities at a handoff point, steps/handoffs that should be emphasized in PSM element/system rollout training (see Section 4.5), and steps that may need oversight/control (and/or data that can be used to measure system performance and health; see Section 4.4.8).

Workflow diagrams can be developed as discussed above, or other techniques can be used, particularly if the facility or organization has personnel available who are familiar with their use. These techniques might include:

- using the "procedural" HAZOP approach to analyze the workflow steps (see the CCPS book *Guidelines for Hazard Evaluation Procedures*) or
- using applicable techniques from business/process improvement systems (such as Six Sigma, TQM, Lean Systems, or Business Process Reengineering) to analyze and improve the workflow.

Finally, if you plan to use any software for managing PSM element activities that has a built-in workflow (common to MOC, incident investigation, and corrective action tracking software), the existing software workflow diagram should be (1) obtained, (2) critiqued as discussed above, and (3) changed (if possible) to address issues and/or improvement opportunities. The final software workflow diagram should be published as part of the new PSM element written program procedures. Training and validation of understanding should be provided as needed.

4.2.2 Example Workflow Diagrams

Examples of workflow diagrams that may result from these efforts are shown in Figures 4.1 through 4.5. Figure 4.1 is a simple, high-level workflow for MOC (shown as Figure D.1 in *Guidelines for the Management of Change for Process Safety* [Ref. 4.6]). Figure 4.2 is a more detailed incident investigation workflow (shown as Figure 19.1 in *Guidelines for Risk Based Process Safety* [Ref. 4.3]). Following these are examples of a more detailed MOC workflow (Figure 4.3), a PHA workflow (Figure 4.4), and a compliance audit workflow (Figure 4.5).

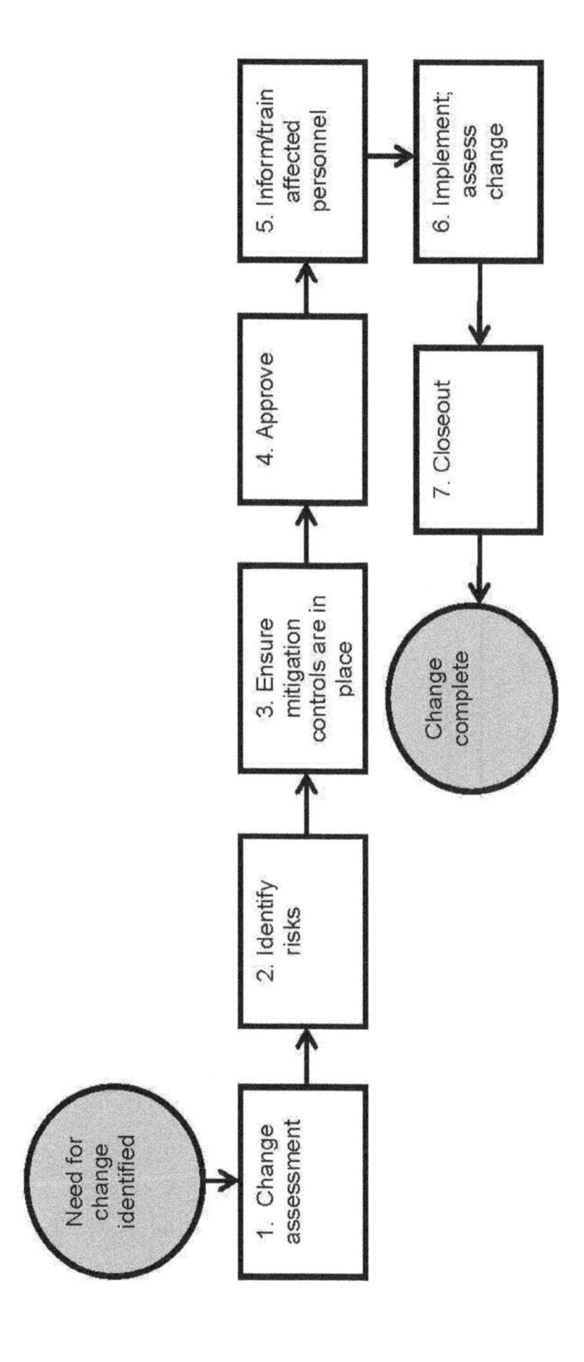

Figure 4.1 Simple Generic MOC Workflow

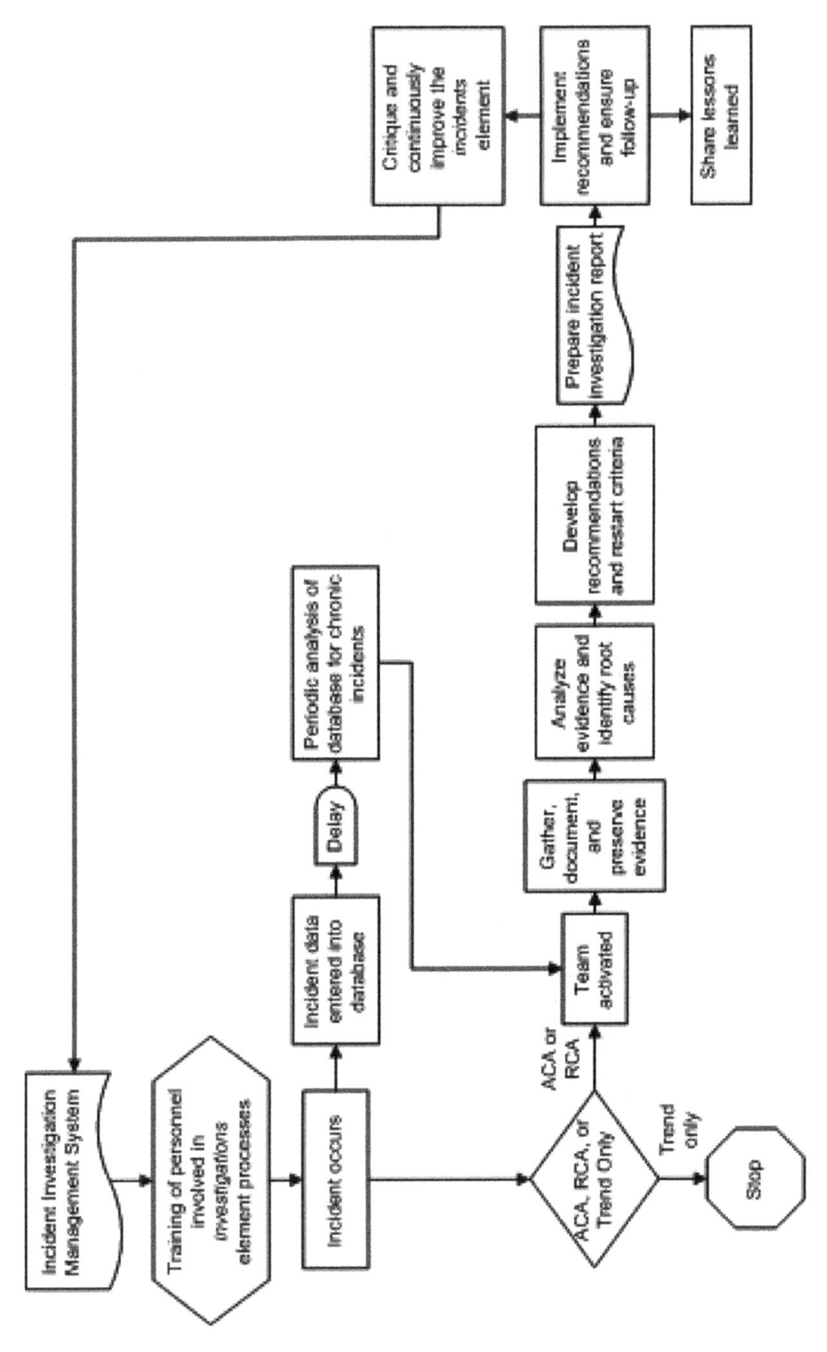

Figure 4.2 Incident Investigation Workflow

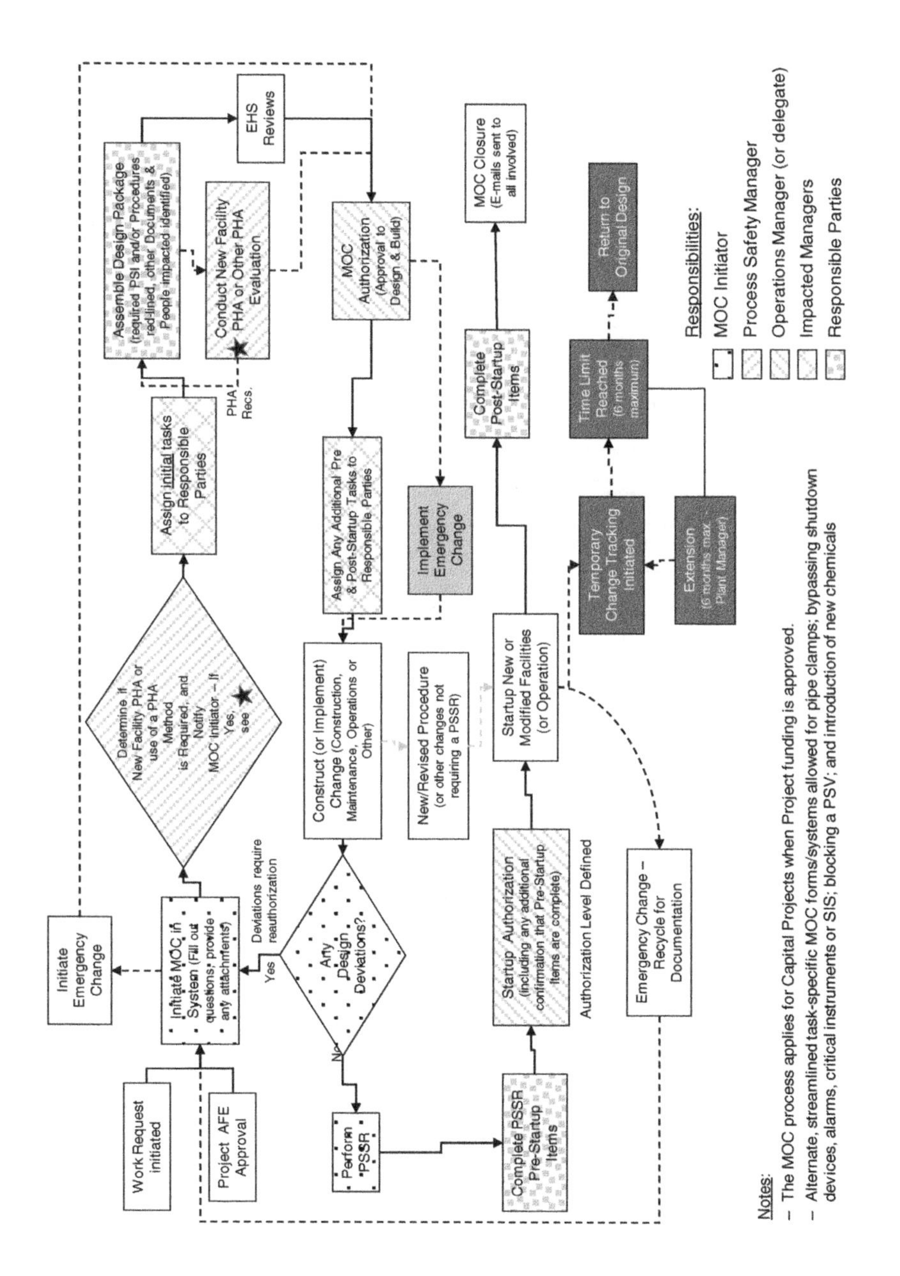

Figure 4.3 Detailed MOC Workflow

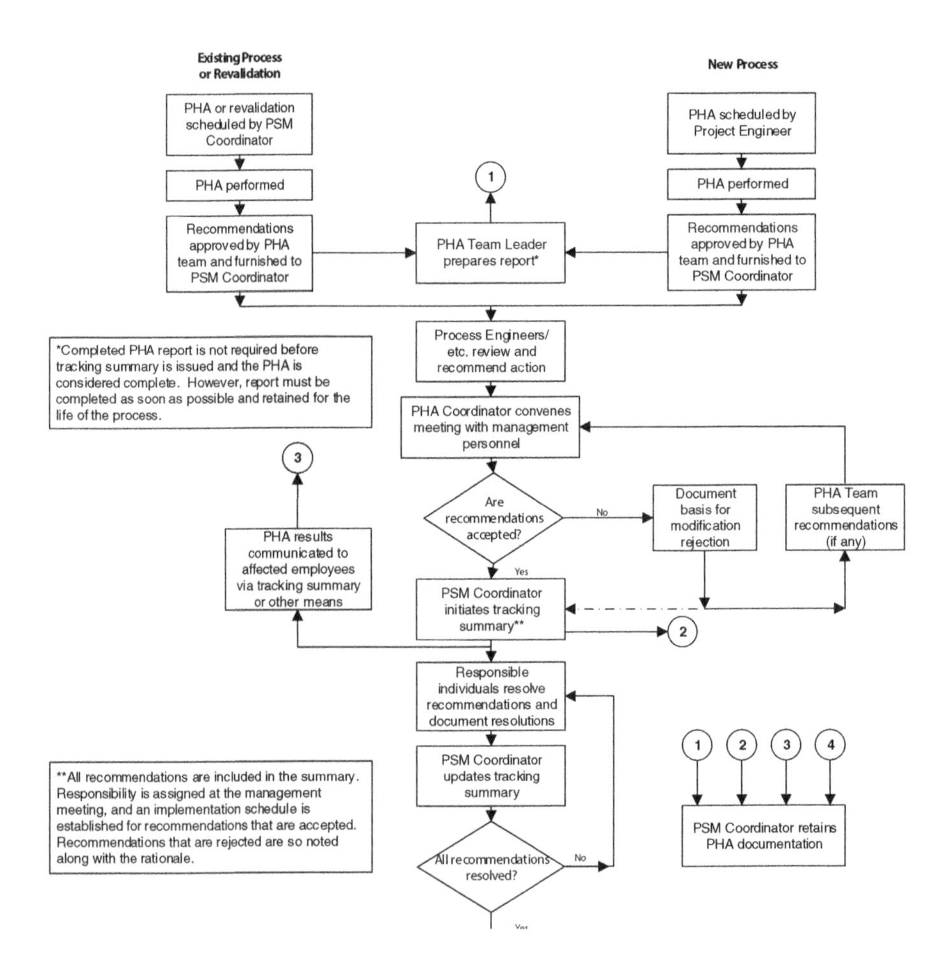

Figure 4.4 PHA Workflow

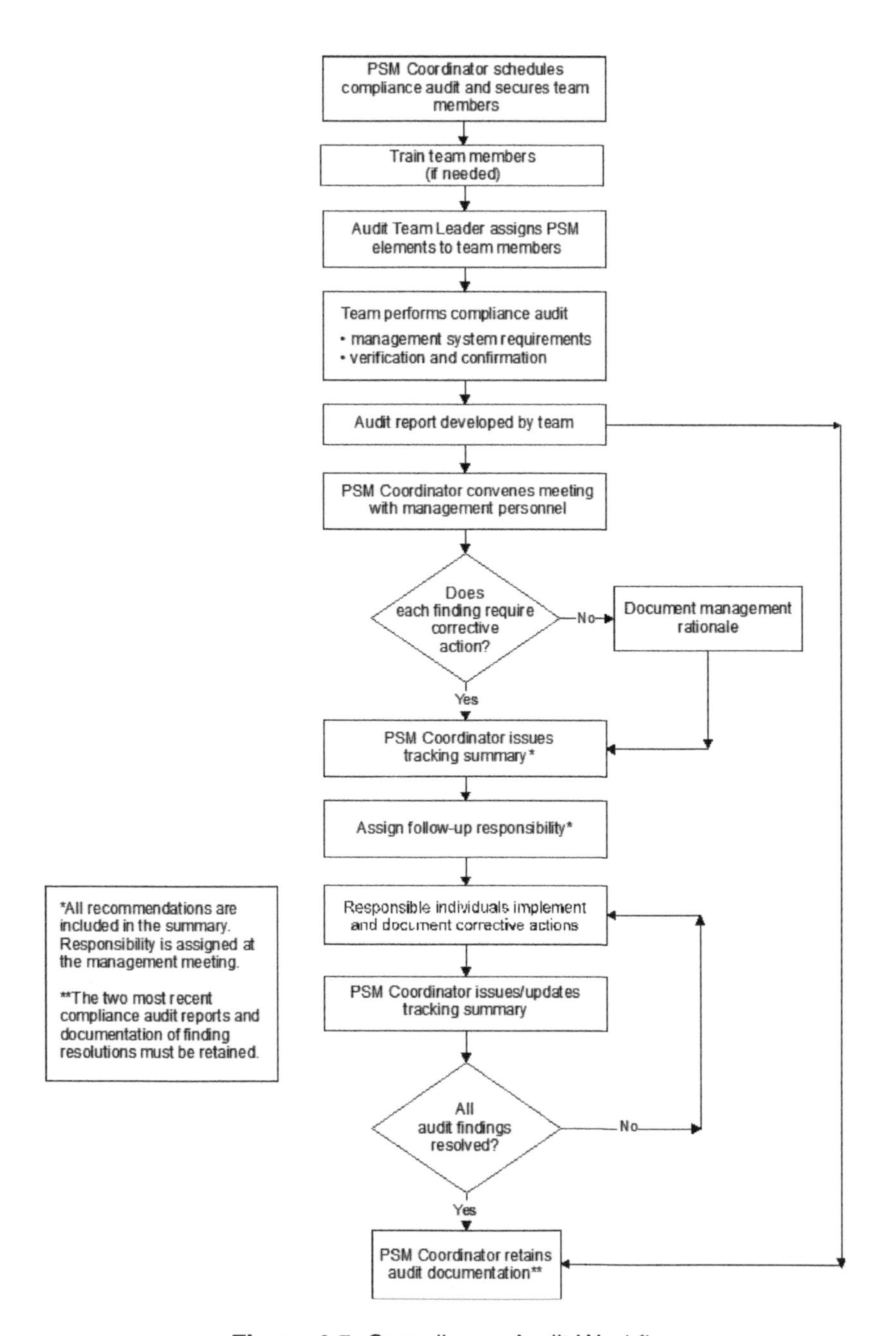

Figure 4.5 Compliance Audit Workflow

4.3 ESTIMATE THE WORKLOADS AND RESOURCES

The next step in implementing a new PSM system is to estimate the element and system workloads and resources. This is important in setting up the new system for success – without an adequate understanding of the workloads and resources that the new elements and the overall system will require, the program is destined for failure. Therefore, the element and system workloads and resources need to be (1) estimated based on their design criteria and workflows, (2) compared to the workloads/resources being expended in the existing systems, (3) compared to the resources available, and (4) adjusted in advance if it is anticipated that inadequate resources are available.

4.3.1 Define the PSM Element/System Design Parameters

Based on the design specifications (see Section 4.1) and workflows (see Section 4.2) that have been developed, the design parameters for each PSM element and the overall system can now be established. These parameters should include:

- lists of initial tasks required to implement the new PSM system,
- lists of existing and new PSM-related activities, and
- the associated existing and new resources required to accomplish these tasks and activities.

4.3.2 Review Possible Sources of PSM Activity and Resources Information

As discussed in Section 4.1, the CCPS RBPS book and books written for individual RBPS/PSM elements provide a wealth of information on (1) possible PSM activities for each element and (2) tools for assisting in performing these activities.

The PSM team (established as discussed in Chapter 3) should now be in position to define the (1) initial (one-time) tasks required, (2) the existing and new PSM activities that will need to be established on an ongoing basis, and (3) the existing and new resources required to accomplish these tasks and activities.

The necessary resources may include (1) more or different types of personnel; (2) consultants/engineers; (3) travel expenses (for assistance from other sites or consultants/engineers); and/or (4) new procedures, forms, or software that have to be developed, purchased, and/or supported.

4.3.3 Estimate the PSM Workloads and Resources

Estimating the workloads and resources for implementing a new PSM system should be done on an element-by-element basis. For some elements, there may be little or no change in workload or resources compared to the existing system. For others, increased workload or resources may be required only for initial implementation and not on an ongoing basis. In either of these cases, documentation of the expected impact or a simple "ballpark" estimate of the initial implementation effort may be adequate.

However, some elements in the new PSM system may require a significant increase in workload and resources. For these, it will be worthwhile spending time and effort on developing more detailed estimates. A simple approach is described below.

Element tasks and interconnections

First, either use the workflow you developed using the guidance in Section 4.2, or brainstorm all of the tasks required to address it, as shown in Table 4.11. You will notice as you complete this exercise that in many cases tasks will be interconnected, so that the product of one task provides the starting point for another task. For example, defining PHA-related training needs depends on defining PHA procedures. These interrelationships, along with the priority elements and facilities you have defined, should help determine the order in which you undertake the tasks.

At another level, PSM element interrelationships must be considered. For example, it will not be possible to install a PHA program until the PSI is up to date. Figure 4.6 illustrates the interrelationship among several RBPS PSM system elements, suggesting the order in which implementation or related tasks might be undertaken.

Table 4.11 Example Work Breakdown Structure: Priority Elements and Tasks

Priority Elements	Tasks
Hazard identification and risk analysis	• Define process information needs • Develop PHA procedure flow • Document PHA procedures • Define staff training needs • Develop follow-up tracking process
Process knowledge management	• Define P&IDs needed • Establish management system for future updates
Training	• Define training requirements • Define training management system • Develop training program • Pilot test training • Implement ongoing program

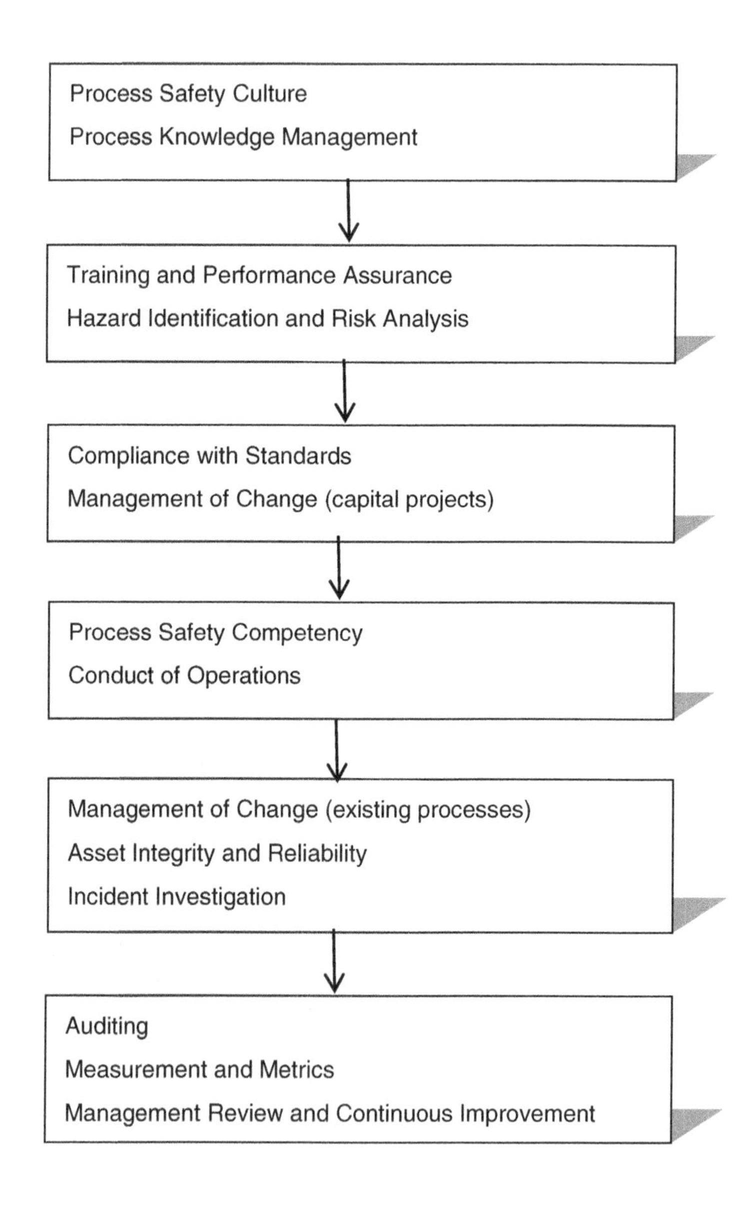

Figure 4.6 Example of Interrelated PSM Elements and Implementation Order

Once these lists and the "network" of tasks have been established, your next step might be to define a work breakdown structure for each task. A work breakdown structure shows the individual steps or elements required to complete each task. Each of these work elements can then be assigned to individuals or groups for action.

Staffing

After the necessary tasks and work products have been identified and risk ranked, consider the skills required to accomplish them and where you might be able to find them. One approach is to develop skills/tasks matrices, indicating what types of skills are required for each of the identified tasks, as shown in Table 4.12. After this is done for each PSM element, the overall skill needs for the complete program can be understood.

Note: It is important to ensure that the personnel selected for any role within the PSM management system, and especially the person leading the implementation or improvement effort, have the proper background, experience, and qualifications to succeed in the role. However, a full discussion of this subject is beyond the scope of this book.

Note: Matching skills with tasks can be assigned to one or two PSM team members rather than requiring the full group's attention. In making such an assignment, keep in mind that this subtask creates the foundation for estimating resources (see the "Resource Plan Development" subsection later in this chapter).

Table 4.12 Example Matrix: Skills and Tasks Required for Developing a PHA Program

	Process Engineering	Process Safety Engineering	Operations	Computer Systems Analysis	Training
Define process information needs	X	X	X	X	
Develop PHA procedure	X	X	X	X	
Document PHA procedure	X	X	X		X
Define training requirements		X	X		X
Develop follow-up tracking process		X	X	X	

Within the context of staffing questions, you will also want to think in terms of supervision or assigning responsibility for various tasks and subtasks. Each task you identify must be assigned to an individual who will assume responsibility for carrying it out in order to ensure accountability. (The full team will probably prefer to be involved in these determinations, since all members have a stake in how responsibilities are divided up.) As a practical matter, you will likely find that certain clusters or groups of interrelated tasks fall together logically, suggesting that they are best supervised by the same individual. (For example, in Table 4.13 the tasks of developing a PHA procedure and documenting the PHA procedure are closely related and are assigned to the same task leader.) At the same time, you need to be sensitive to the demands on team members' time, as well as that of colleagues the team proposes for task leadership roles.

Note: It may be useful to consider the PSM implementation plan, at this stage of its development, as the site or company "ideal" within the limitations you have established for its scope. By first identifying what is needed, independent of the constraints of time or resources, you emphasize the tasks themselves as the substance of the plan – the core that drives decisions about resource allocations, rather than the other way around (for an example, see Table 4.14). The result may prompt your team to think more creatively about schedule and resource requirements, as discussed in the following sections.

Schedule development

Any schedule you develop should reflect both the tasks you have defined and the resources available for accomplishing them. In addition, the implementation schedule in some cases may depend on a predetermined end date. For this reason, schedule and resource requirement tasks should be seen as interdependent; it is realistic to expect that you may need multiple iterations before both are firmly established.

Table 4.13 Example Task Leadership Assignments for PHA Program Development

Task	Responsibility
Define process information needs	Program Manager
Develop PHA procedure	Process Safety Engineer
Document PHA procedure	Process Safety Engineer
Define training requirements	Training Specialist

Table 4.14 PSM Staffing Needs by Element and Function

	Staff (Weeks)				
PSM Element	Process Engineering	Process Safety Engineering	Operations	Computer Systems Analysis	Training Specialists
Process Safety Culture	2	10	10	0	10
Compliance with Standards	4	6	2	1	2
Process Safety Competency	6	6	4	0	2
Workforce Involvement	0	2	2	0	1
Stakeholder Outreach	0	2	0	0	0
Process Knowledge Management	24	8	8	20	2
Asset Integrity and Reliability	30	12	30	18	2
Contractor Management	4	4	4	2	4
Training and Performance Assurance	4	4	4	4	30
Management of Change	12	8	26	12	2
Operational Readiness	4	4	12	2	2
Conduct of Operations	0	4	12	2	2
Emergency Management	1	4	4	1	2
Hazard Identification and Risk Analysis	10	18	4	4	6
Operating Procedures	4	16	22	4	10
Safe Work Practices	0	4	8	2	4
Incident Investigation	4	12	6	0	2
Measurement and Metrics	1	4	1	2	0
Auditing	1	34	4	2	4
Management Review and Continuous Improvement	0	4	2	1	0
TOTAL	111	166	165	77	87

Each of the tasks and subtasks that make up the work breakdown structure should be relatively simple to estimate in terms of time required. If no one on the PSM team has specific experience with a given task, find someone who does (a facility manager, for example) and ask for his or her input. Once the time requirements for each task have allowed you to define the overall time required for each element, these estimates can be combined to provide a total implementation plan.

In addition to specific deadlines for each task and subtask, you should also consider implementation milestones (i.e., key points in the plan at which you will want to review the progress to date and make any necessary adjustments). For example, if you have adopted an element-by-element approach, you could consider completion of each PSM element as a milestone. Similarly, each facility's program would constitute a milestone, if that is the approach you have chosen.

Note: In setting schedules and deadlines, try to be realistic while maintaining your sense of urgency. Be sure to build in sufficient time for group meetings to evaluate the progress to date. Also, remember to consult those most immediately affected by the plan (for example, people whose staff will be called upon). Their existing priorities must be considered during the scheduling process.

The final plan schedule should be presented in an appropriate format, such as a linear timeline (as illustrated in Figure 4.7 for a single element and in Figure 4.8 for the overall program, if you are implementing the selected RBPS PSM system elements shown). Your plan may call for multiple tasks to be undertaken simultaneously. This should be clearly indicated.

An example of an even more detailed "project plan" for PSM implementation is provided in the files on the Web accompanying these guidelines.

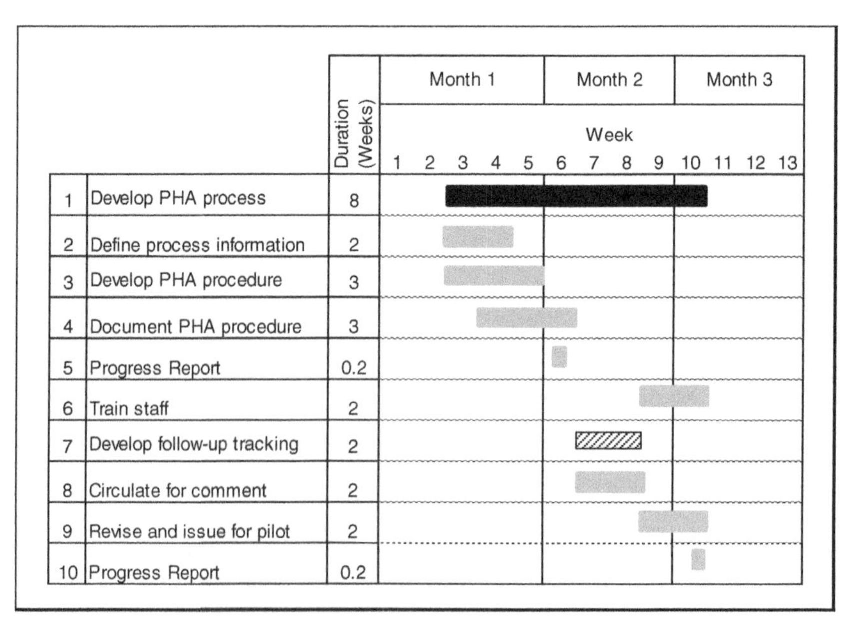

Figure 4.7 Final Plan Schedule – Linear Timeline

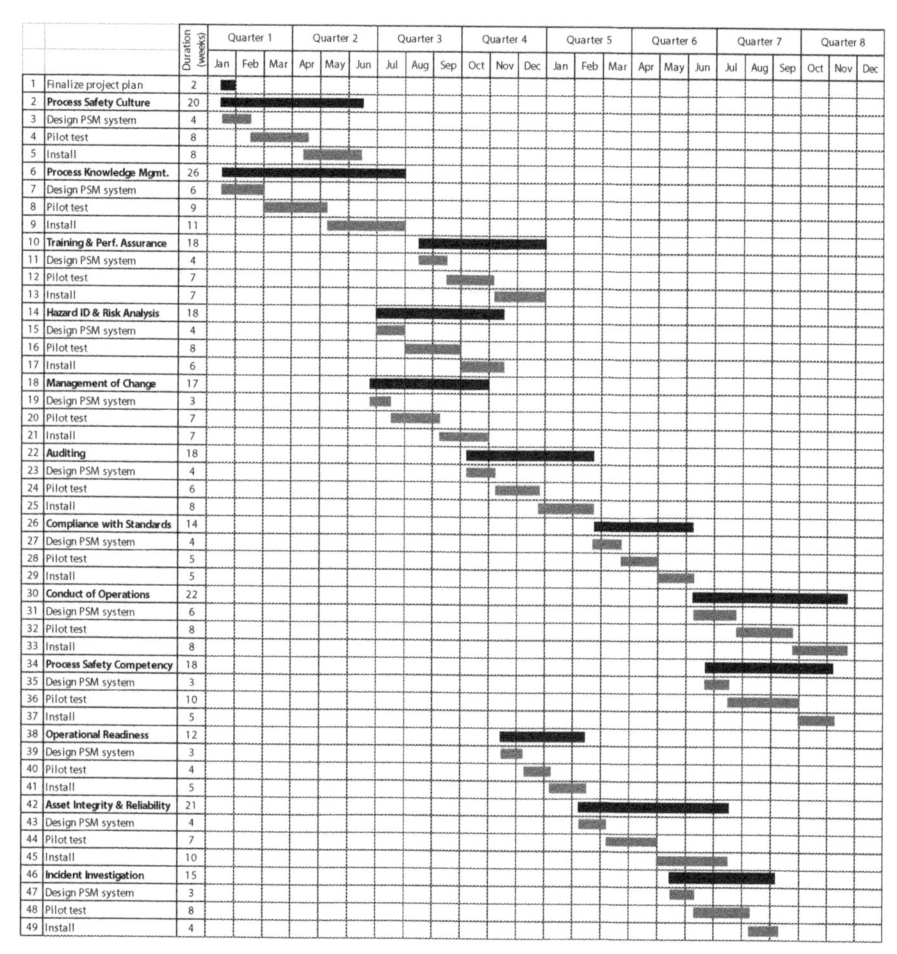

Figure 4.8 Final Plan Schedule for Overall Program

Resource plan development

Most of the basic resources needed are fairly self-evident; staff time will almost certainly be the largest single cost. Support expenses and travel also require funding. In addition, in the course of your work to date you may have identified specific resource requirements, such as computer software for hazard analysis or project management, or consulting services that fill in specific gaps in the knowledge base.

Human resources

The skills matrix developed as part of the program plan forms the basis of your resource requirements estimate. In addition to the skill sets identified for each task, you will need to estimate the level of effort required at each stage of the work breakdown structure (see Table 4.15). As with the scheduling exercise, team members should actively participate in this discussion, since they reflect varying areas of experience. Consider these requirements in as much detail as possible; for example, consider the number of staff-days of effort required to redraft piping and instrumentation diagrams (P&IDs), or the number of days of operations engineering time needed to prepare operating procedures. The sum of individual task resource requirements provides an estimate of overall project needs.

Table 4.15 Staff Resources Plan for "Develop PHA Procedure"

	Staff-days Required							
	Define Process Information Needs	**Develop PHA Procedure**	**Document PHA Procedure**	**Define Training Needs**	**Develop Training Process**	**Revise Per Comments**	**Develop Progress Reports**	**TOTAL**
Program Manager	3	8	2	2	8	8	4	35
Process Safety Engineer	3	8	8	2	2	3	0	26
PHA Leader	3	8	8	0	4	3	0	26
Training Specialist	3	8	0	8	4	8	0	31
TOTAL	**12**	**32**	**18**	**12**	**18**	**22**	**4**	**118**

As a next step, identify the specific resources that are currently available and try to quantify them in terms of staff time. Also, be sure to factor in any restrictions that might apply (e.g., a company policy that limits reassignment of support staff or confines travel to a specific staff level). Ask each PSM team member to participate in this exercise, since each may have a different perspective and/or access to different resources within the company.

Then compare the resource list with the staff allocation estimate, as shown in Table 4.16. The gaps that emerge from this comparison represent areas for the team to examine closely and determine how best to address them. Some of the questions the team should ask include:

- Do alternative resources exist?
- Is there another way to accomplish this task?
- Would a schedule adjustment free up needed resources?
- Should we rethink the priority assigned to this task?

If the answer in each case is "no," the team must consider other options and their cost. Some of these might include:

- reassigning resources from another, lower priority task;
- hiring additional project staff; and/or
- retaining consultants.

Table 4.16 Example of Ideal Staff Needs vs. Actual Availability Analysis

	Staff-weeks							
	Quarter 1		Quarter 2		Quarter 3		Quarter 4	
	Needed	Available	Needed	Available	Needed	Available	Needed	Available
Process engineering	24	20	24	26	16	26	15	20
Process safety engineering	16	26	23	26	24	26	24	39
Operations	13	13	21	26	23	13	21	13
Computer analysts	12	9	12	10	7	9	6	9
Training specialists	12	9	20	13	4	13	0	9

Other resource requirements

As you review each of the tasks, you should also consider what other resources (other than staff time) may be needed. For example, improvements to incident/near-miss reporting might require improved network communications between facilities and headquarters, or a particular training module could be purchased to address an identified gap.

For those areas where you think outside services will be required, you should get estimates from qualified consultants.

In addition, working from the schedule you have developed, you will need to estimate travel costs. Key dates to keep in mind include those you have established as milestones, since they may call for group meetings to review progress.

Once you have estimated the level of effort required for each task, you can, with some reliability, estimate project support requirements (secretarial, administrative, etc.) at 15 percent to 20 percent of professional staff time.

Identification of variables and potential issues

Even the best-laid plans are subject to last-minute changes and unexpected variables, and there is an element of uncertainty in every estimate. Some of the more common sources of variance include factors beyond your control, such as:

- economic setbacks affecting available resources,
- price increases for major out-of-pocket items,
- layoffs,
- business strategy shifts that change key processes, and
- emergencies that command immediate priority.

Other variances arise from causes you can more readily predict, such as:

- tasks that turn out to be far more (or less) complex than estimated,
- necessary skill levels that are higher (or lower) than estimated,
- available skill levels that are lower (or higher) than estimated,
- an increase (or decrease) in required travel, and
- reporting that is more (or less) time-consuming than expected.

While none of these variances can be precisely estimated, you should consider them (and others you identify) in terms of their potential impact, and build appropriate contingency percentages into estimates of both time and resources. In addition, where you know estimates to be soft, identify them as such; this is preferable to creating unrealistic expectations.

As a final task in estimating resource requirements, ask the team to challenge the estimates by thinking of ways to disprove the data. Given the range of experience team members represent, they should be able to poke holes in even the most precise, well-considered estimates. This will help you identify areas in which your numbers may be soft and flag those in which you may have been overly ambitious in terms of time requirements. This exercise will also help when you present the estimate to management, as you can anticipate their questions.

Once you have developed the implementation plan and schedule and estimated resource requirements, you should organize this information into a manageable format for presentation to your PSM sponsor. If you have kept management informed about your progress, you should have a clear idea of what is expected in terms of format, content, and level of detail. A sample table of contents for a project plan is shown in Table 4.17, and Table 4.18 provides additional detail on selected portions of the plan.

4.4 DEVELOP WRITTEN PROGRAMS/PROCEDURES

The CCPS RBPS book defines a management system as "a formally established set of activities designed to produce specific results in a consistent manner on a sustainable basis." A PSM management system must be described in written programs and procedures to be sustainable: unwritten practices will fade away as time passes, key personnel leave, or challenges arise that cause "temporary" deviations from the system, which become accepted and eventually permanent.

Note: Although regulations may require "written programs" for only a few elements (e.g., the OSHA PSM regulation only requires them for employee participation and MOC), it is virtually impossible to establish a sustainable management system without a written program.

4.4.1 Purpose and Scope

Written programs for PSM elements and the overall system should include both a purpose (i.e., the objectives that the element/system is intended to achieve) and a scope (i.e., where the element/system must be applied and followed).

Stating the purpose helps those working on the element/system understand the benefits of their efforts, thereby improving the buy-in and alignment within the organization. A clear purpose also helps the organization align on the scope (for example, the extent to which the element will be applied to noncovered processes and equipment).

The scope of application of the element/system should consider the following:

- The minimum requirements based on PSM regulations, if applicable. For example, the element (or portions of it) may apply to certain chemicals and/or processes in the facility.

- RBPS considerations of the hazards and level of risk involved with each chemical/process
- The benefits of applying the element to noncovered processes or portions of the facility. For example, many companies apply elements such as MI, MOC, and incident investigation to all the chemical processes in their facilities because they believe that the level of risk and/or cost reduction achieved offsets the additional effort required.

A clear scope also avoids internal confusion about the applicability of the PSM requirements and guidelines.

Table 4.17 Example Project Plan Table of Contents

1. Organization and responsibilities
    - 1.1 Project description
    - 1.2 Project organization
    - 1.3 Project responsibilities
2. Administration
    - 2.1 Communication
    - 2.2 Documentation
    - 2.3 Progress reporting
3. Project controls
    - 3.1 Work breakdown structure *(see Table 4.11)*
    - 3.2 Schedule *(see Figure 4.8)*
    - 3.3 Resources needed *(see Table 4.15)*
    - 3.4 Resource estimates *(see Table 4.16)*
4. Quality plan
    - 4.1 Quality management objectives
    - 4.2 Project quality assurance
5. Installation and verification *(see Table 4.18)*
    - 5.1 Pilot testing
    - 5.2 Training
    - 5.3 Installation
    - 5.4 Verification
    - 5.5 Financial controls

Table 4.18 Example Installation and Verification Plan

Installation and verification
Pilot testing. Pilot testing of each PSM system will be performed at two locations to be selected by the project team. The test period will be two months. During the test period, the project team will monitor the PSM system to ensure that procedures are clear and do not conflict with other procedures.
Training. The project team will define the training needed for implementation of each new PSM system. The amount and type of training, and the number of staff requiring training, will be identified in the procedure developed.
Installation. Installation of each new PSM system will occur after the pilot test phase. Documentation on the procedure will be circulated, and the procedure will be described at a meeting of the company Engineering Council.
Verification. Each PSM system will include a mechanism for verifying installation. This will either be the requirement for copies of documentation being sent to Corporate EHS, or another mechanism defined by the project team.
Financial controls. All effort expended in support of this project will be tracked by the Controller's office and reported to the Project Manager. Accumulated costs will be reported in monthly project team progress reports *(see the example shown in Figure 4.9).*

4.4.2 Personnel Roles and Responsibilities

Identifying personnel roles and responsibilities for all PSM elements and activities is important. Problems can arise when (1) personnel believe something is someone else's job or (2) the specific responsibilities are not clearly spelled out (as the saying goes, "When it's everybody's job, it's usually nobody's job"). In addition, important oversight and continuous improvement activities (e.g., initial reviews, authorizations, auditing, metrics, management reviews) can easily be overlooked if they are not defined and documented.

Developing workflows for elements/systems (see Section 4.2) naturally includes a discussion of "who does what"; so if there is a workflow, use it as a starting point for defining roles and responsibilities.

Another common practice to clarify personnel roles and responsibilities is to develop RASCI (Responsible, Accountable, Supports, Consulted, Informed) or RACI charts for each element. Such charts identify (1) all the tasks related to the element that warrant attention, (2) the personnel who should be involved in each task, and (3) their roles for each task. In addition, having these charts available can allow the tailoring of training specific to people's roles within the organization. Figure 4.10 provides an example MOC RASCI chart.

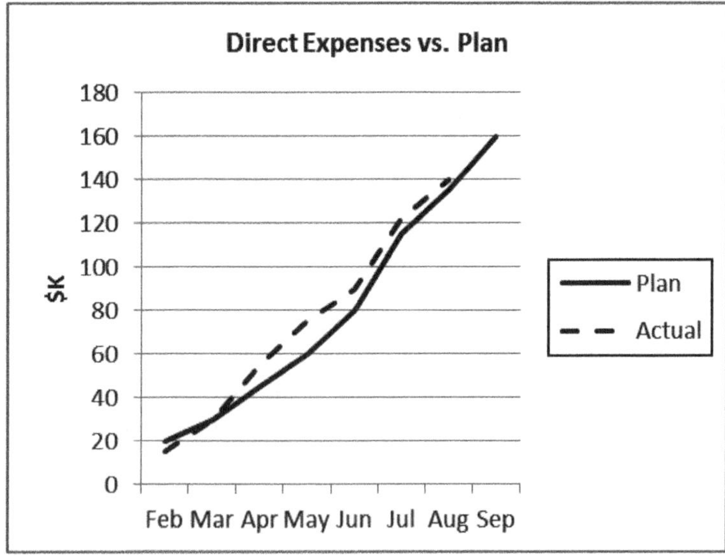

Figure 4.9 Sample Reports: Resources Used and Expenses vs. Plan

R – Responsible *(the person doing the work)*
A – Accountable *(generally supervises the person doing the work and is ultimately accountable)*
S – Supports *(provides information, resources, or assistance)*
C – Consulted *(has a vote or say about the work)*
I – Informed *(needs to know about the work)*

Task	*MOC Initiators*	*Impacted Managers*	*Responsible Parties*	*Operations Manager (or delegate)*	*Responsible Subject Matter Expert*	*EHS Trainer*	*Site Process Safety Managers*	*Plant Managers*	*Corp. Operations VP*	*Corp. EHSS Director*
Initiate MOCs in the MOC application	R							A		
Review implemented MOCs for any Design Deviations	R							A		
Lead PSSRs	R							A		
Review MOCs for Impact & Assign Initial Tasks		R, A								
Assign Additional Pre- or Post-Startup Tasks (after MOC Authorization)		R, A								
Complete MOC Reviews & Assemble Design Package			R	A						
Complete Assigned Pre-& Post-Startup Items			R					A		
Review MOC Design Package & Authorize MOC				R, A			C			
Review MOC/PSSR Documentation & Authorize Startup				R, A			C			
Determine need for PHA or Method, Ensure PHAs Meet Requirements							R			A
Implements/Maintains Program/Procedure					R	C	C		S, A	S
Determines training requirements					R	S	S	A		
Updates Training Matrix					I	R	I	A		
Coordinates with other PSM element or EHS programs							C		R	A
Provides metrics to Site Management					R		C	A		
Review/approve procedure revisions					R		R	R	R	R, A

Figure 4.10 Example RASCI Chart for Management of Change

- A "Who Does It?" section for each RBPS element that provides a high-level view of the personnel involved in element activities. This should serve as a good starting point for any site/company.
- A "Possible Work Activities" section for each element, which should be reviewed as discussed in Section 4.1 of this book. This should lead to a discussion of what groups and personnel should be involved in each activity that your facility decides to adopt.
- Similarly, the "Examples of Ways to Improve Effectiveness" section should lead to a discussion of roles and responsibilities regarding the activities selected. For example, some of the tasks in this section of the "Operating Procedures" chapter of the RBPS book for which roles and responsibilities would need to be determined and defined are:
    - periodically auditing conformance to procedures,
    - eliminating shift-to-shift differences, and
    - simultaneously reviewing procedures for multiple facilities operating the same process.

4.4.3 Work Processes, Tasks, and Procedures

Your written PSM programs/procedures should include the following:

- **The work process(es) to be used in executing element activities.** If a workflow has been developed per Section 4.2, it should be an integral part of the procedure. Even if the work process does not warrant a workflow diagram, the procedures should lay out the steps involved, the order of the steps (when important), and any handoffs between personnel and/or workgroups.
- **The defined tasks/activities for the element.** Section 4.2 describes one process for determining the "design specification" for each element, down to the task/activity level. In any case, these should be delineated in the written programs/procedures in order to (1) ensure that the organization knows everything that is to be done and (2) "institutionalize" them by ensuring that the program and the practice are the same.
- **Adequate details to understand the requirements and guidelines for the element activities.** Keeping in mind that you want expectations to be clear, consistent among workgroups, and consistent over time, it is important to clearly describe them so that everyone's understanding and practice are essentially the same.

4.4.4 Necessary Inputs and Anticipated Results

Written programs/procedures should also address the necessary inputs (to the element as well as to tasks/activities within the element) and the anticipated results (overall and for each task/activity).

If a workflow (see Section 4.2) has been developed, it will provide a starting point for inputs and outputs for each work process and for many of the steps within it. In addition, the RBPS book element chapters include a section entitled "What is the Anticipated Work Product?" that should provide guidance on the outputs and anticipated results for the overall element. Based on a review of this and any other available information, and on appropriate discussions with stakeholders, the necessary inputs and outputs should be determined and documented in the written program. The documented inputs and outputs provide a starting point for potential measures and metrics to verify the health and effectiveness of the element. The quality or completeness of the inputs and outputs, if measureable, provides a tool to ensure the performance of the system (see Section 4.4.8).

4.4.5 Personnel Qualifications and Training

Without qualified, trained personnel involved in executing the element activities, the element/system is likely to fail. Therefore, the written programs/procedures need to address the qualifications and training required, in general and/or for specific tasks/activities within each element.

Once again, the RBPS book can be very helpful, as the standard "Maintain a Dependable Practice" section (especially within the "Possible Work Activities" section in each chapter) provides guidance on what is needed in order to involve competent personnel. It suggests the specific training these personnel should receive and/or experience. In addition, the "Process Safety Competency" element (Chapter 5) provides extensive guidance on ensuring competency, both in the overall organization and in the execution of each PSM element. Based on this information, your site's or company's experience, and the existing level of competency, determine what PSM activities within each element require qualified personnel and define the training and/or experience required to achieve and sustain qualification.

4.4.6 Activity Triggers, Schedules, and Targets

PSM activities may be triggered in many different ways. For example, they may be based on:

- a frequency (e.g., annual certification of operating procedures, compliance audits that may be required every 3 years, PHA revalidations that may be required every 5 years, some pressure vessel internal inspections that may be scheduled every 10 years, periodic refresher training),
- a change (e.g., a new section of piping is to be added or an operating procedure is to be significantly revised; therefore, an MOC is initiated, subsequently leading to PSI, operating procedure, and MI program updates, as well as a pre-startup safety review [PSSR]),

- a planned event (e.g., a unit is taken down for turnaround maintenance so the "shutdown for maintenance" procedure is executed, followed by a number of safe work practices and maintenance procedures, after which the "startup after turnaround" procedure is performed), or
- an unplanned event (e.g., a release of a hazardous chemical occurs, leading to an immediate emergency response and a subsequent incident investigation).

The design specification (see Section 4.1) should have identified most if not all of the triggers (or initiating events) applicable to each element. The "Where/When is it Done?" section in each RBPS book element chapter also provides guidance on this subject and is worth reviewing.

All activity triggers for an element or for the activities within it should be determined and documented in the written programs/procedures.

Schedules for PSM element activities can be very valuable, especially when large and/or infrequent activities must be performed on a set frequency to comply with regulatory requirements. For example, a site with 25 PHAs, each of which must be revalidated every 5 years, needs a documented schedule in order to avoid becoming overdue. Schedules may be documented/maintained in a table that is periodically updated, laid out as a "Compliance Calendar," or maintained in a computer system (e.g., training management, maintenance management, or enterprise asset management systems).

The written programs/procedures should either include applicable schedules or clearly reference where/how they are maintained.

Target dates may apply within a PSM system with regard to meeting a frequency-based schedule, as previously discussed. However, they should also be considered with regard to establishing limits on the time required for certain activities. Here are some examples:

- Setting target dates for the implementation of corrective actions resulting from PHAs, incident investigations, compliance audits, etc. Examples are provided throughout the RBPS book.
- The time for which a temporary MOC is allowed to remain in place may be limited to a few months in order to ensure that it is not "forgotten" and the associated risk is limited.
- The time for which an engineered pipe clamp is allowed to remain in place may be limited to a few years in order to ensure that the associated risk is limited and such actions do not become too commonplace and accepted (a "normalization of deviance").
- A site may require that all PSI associated with an MOC be updated within 90 days after the PSSR is completed in order to ensure discipline in completing this "final" step.

- A site may require issuance of incident investigation reports within 30 days of the incidents in order to (1) emphasize the importance of prompt and thorough investigation and (2) support learning from incidents.

After thorough consideration of which element activities should incorporate targets, these should be documented in the written programs/procedures and used as a measure/metric as appropriate.

4.4.7 Resources and Tools Needed

Development of the design specifications for each element and the system (see Section 4.1) and resource estimates (see Section 4.3) should have identified the resources and tools needed for successful implementation. In the written programs/procedures, any specific resources/tools that are to be utilized should be documented. Examples include specific (1) forms or electronic documents, (2) hard-copy or electronic files, (3) equipment (e.g., for MI ultrasonic thickness or vibration checks), and (4) software for PSM activities (e.g., document management systems, PHAs, MOCs, incident investigations, MI inspections/tests, corrective actions). In addition, all references that are to be utilized or that support the program should be cited (e.g., regulations, corporate standards, RAGAGEPs).

Note: Appendix III ("RBPS Tools"), and the "Eli Lilly and Company PSM Tools" and "PSM Software Compilation" on the files on the Web accompanying this book, provide information on some resources/tools that may be useful.

4.4.8 Measurement, Management Review, and Continuous Improvement

Serious process safety incidents occur relatively infrequently; but when they do occur, they usually involve a confluence of root causes, some of which involve degraded effectiveness of management systems or, worse, complete failure of management system activities. Facilities should monitor the real-time performance of management system activities rather than wait for accidents to happen or for infrequent audits to identify latent management system failures. Real time monitoring will enable the identification and correction of abnormalities before a serious incident occurs.

In the RBPS management system, the "Measurement and Metrics" element establishes performance and efficiency indicators to monitor the near-real-time effectiveness of the RBPS management system and its elements and work activities. It addresses which indicators to consider, how often to collect data, and what to do with the information to help ensure responsive, effective RBPS management system operation. Also, this chapter emphasizes that a combination of leading and lagging indicators is often the best way to provide a complete picture of process safety effectiveness.

One or more metrics can be established for each RBPS element, or a few can be created for the entire system. Metrics can address performance issues, efficiency issues, or both (effectiveness) in all operating phases. Once data-gathering/refreshing systems are in place, metrics can ideally be viewed anywhere. The frequency for refreshing the individual metrics can range from daily to weekly to monthly or longer, depending upon the importance/value of the metric, the dynamic nature of the metrics, the anticipated costs of data collection, and the local needs.

Establishing metrics is simpler to do during the initial design and implementation of a new PSM system than after the system has been installed and made operational. Each RBPS element chapter has a section that contains a list of possible metrics proposed for that element's key principles. Readers can select from these examples or develop and add their own metrics (see Section 4.4.4 for information on developing metrics from inputs and outputs.). Typically, a small set of metrics is proposed, data are gathered, and the set is pilot tested to determine whether tracking the metric data helps identify management system degradation. There may be a different set of metrics for the initial implementation to verify that the system is meeting the design. The collection and use of metrics may also be changed or used in a rotation to balance the use of resources necessary to collect, monitor, and assess against the value of a wider/larger metric set. Developing metric strategies will optimize resources and achieve the advantages of a larger metric set.

Other CCPS sources on process safety metrics include:

- the metrics Web site at www.aiche.org/ccps/search/metrics, which provides a wealth of information, including brochures, presentations, and Webinars on this subject;
- the 40-plus-page brochure entitled *Process Safety Leading and Lagging Indicators . . . You Don't Improve What You Don't Measure*, available at www.aiche.org/sites/default/files/docs/pages/metrics%20english%20updated.pdf; and
- the *Guidelines for Process Safety Metrics* (Ref. 4.7), which provides basic information on process safety performance indicators, including a comprehensive list of metrics for measuring performance and examples as to how they can be successfully applied over both the short and long term.

Similarly, ANSI/API RP 754 provides possible metrics, grouped into four tiers (Ref. 4.8).

After considering the available information on measurement and metrics, as well as any existing metrics, the facility or company should decide how to measure element/system performance, establish metrics (for elements and/or the overall system), establish metrics strategies to optimize resources, and document all of this in the written programs/procedures.

"Management Review and Continuous Improvement" is Chapter 22 in the RBPS book. The management review process provides regular checkups on the health of process safety management systems in order to identify and correct any current or incipient deficiencies before they might be revealed by an audit or incident, thereby supporting system sustainability and continuous system improvement.

Management review is the routine evaluation of whether management systems are performing as intended and producing the desired results as efficiently as possible. It is the ongoing "due diligence" review by management that fills the gap between day-to-day work activities and periodic formal audits.

Management reviews should be conducted wherever RBPS elements are implemented. While they can be scheduled on an as-needed basis, management reviews of a particular RBPS element are typically conducted at a predetermined interval (e.g., frequencies ranging from monthly to annually are common), and they may be scheduled in conjunction with other regularly scheduled meetings, such as facility safety committee meetings.

Management reviews are conducted with the same underlying intent as an audit – to evaluate the effectiveness of the implementation of an entire RBPS element or a particular element task. However, because the objective of a management review is to spot current or incipient deficiencies, the reviews are more broadly focused and more frequent than audits, and they are typically conducted in a less formal manner.

Nevertheless, like an audit, a management review at least checks the implementation status of one or more RBPS elements against established requirements. The management review team meets with the individuals responsible for managing and executing the subject element to (1) present program documentation and implementation records, (2) offer direct observations of conditions and activities, and (3) answer questions about program activities.

Recommendations for addressing any existing or anticipated performance gaps or inefficiencies are proposed, and responsibilities and schedules for addressing the recommendations are assigned. Typically, the same system used to track corrective actions from audit findings is used to track management review recommendations to their resolution.

After considering the available information on management reviews, the facility should institutionalize the requirements for these in its written programs/procedures.

4.4.9 Auditing

Beyond the "compliance auditing" element that is part of most PSM management systems (see *Guidelines for Auditing Process Safety Management Systems*, *2nd Edition* [Ref. 4.9], for more information on this subject), facilities should consider monitoring their PSM system maturation over time. While audits, in their narrowest sense, focus on verifying conformance with established standards for the implementation of process safety, many organizations will aspire to levels of performance beyond regulatory compliance. Comprehensive implementation of an auditing element helps organizations create a system for monitoring performance over time, which allows them to track the maturation of the management system. Chapter 21 of the RBPS book provides guidance and direction on how to accomplish this. Some key points include the following:

- Relevant performance metrics should be identified for each RBPS element and updated for each audit.
- Where performance is particularly problematic, more frequent or specially targeted audits may be appropriate.
- Results should be trended to determine whether performance is improving.
- Continued poor performance is an indication that weakness exists in the program or support for implementing that particular RBPS element, or that the element is not being consistently implemented in accordance with the established program.
- A root cause analysis may be required to determine the underlying causes of continuing performance problems, including, if pertinent, process safety culture weakness.

Note: For facilities required to complete and maintain periodic regulatory compliance audits, it may be desirable to develop and manage separate reports for audits (assessments) that are not required by regulations.

With all of this in mind, consider what audits will be performed, with what scope (e.g., on individual elements or the whole system), by whom, how often, etc., and document these requirements in written programs/procedures.

4.5 ROLL OUT THE ELEMENTS AND SYSTEM

The next major step in implementing a new PSM system is to roll it out. Activities within this step should include:

- gathering input on how the system should be implemented,
- pilot testing the new system or selected elements,
- developing an implementation plan,
- confirming that the associated tools are ready and the resources are available,
- completing the PSM system procedures, and
- providing rollout training.

4.5.1 Gather Implementation Input

The first steps in rolling out a new PSM system should be to (1) gather input from all the stakeholders on how it should be implemented and (2) learn from previous experiences in the facility or in similar organizations. With these objectives in mind:

- solicit implementation ideas from the workforce, PSM element/system owners, and champions/influencers in the organization;
- solicit implementation ideas from the various departments that will be affected by PSM system implementation;
- consider lessons learned from previous incidents at your company or others, both technical and cultural (see Chapters 19 and 3 in the CCPS RBPS book for more information); and
- collect feedback on how the existing system or elements were previously rolled out and consider what changes these experiences may suggest.

4.5.2 Conduct Pilot Testing

Next, consider pilot testing the new PSM system or selected elements in the field prior to its official implementation. Debugging the proposed PSM system via early pilot testing will provide a better chance for acceptance and success. Normally, PSM programs must be customized to effectively meet the specific needs of the facility or company and its culture. In most cases, the best approach is to perform pilot testing of key, larger elements in phases. This allows the systems to be tested a few times and adjusted as needed before rollout to the entire facility. Finally, build any pilot testing into the implementation plan and schedule.

4.5.3 Develop an Implementation Plan

Select an implementation strategy

There is no single "right" strategy for implementing PSM systems. Depending on your company's needs and culture (which you have defined through your work to date), you may select a strategy that will implement PSM systems company-wide, one that is facility-specific, or a "hybrid" implementation strategy that combines characteristics of both.

Choosing the best implementation strategy for your company depends on a number of factors, most of which you have already identified. As you consider the benefits of each implementation strategy described in this section, it's useful to keep in mind some specific factors:

- The current performance of PSM systems may vary dramatically, either within business segments or among individual facilities within the company.

- The requirements for PSM systems may differ at each manufacturing location because of variations in equipment or local regulation.
- Different manufacturing processes may have different PSM system requirements.
- Company management may be highly centralized or very localized.

Regardless of the implementation strategy you select, try to build in at least some degree of local involvement in planning and implementation. Local involvement can give facility personnel "ownership" of the process, meaning that they will be more likely to work to overcome any problems or false starts. In addition, local participation means that you will be able to identify any local barriers to implementation sooner rather than later.

In all cases, the preliminary work described earlier in this book provides most of the information you will need to select and develop an implementation plan.

Company-wide approach

In the course of your work to date, you have likely identified your company's management style in different divisions, regions, and locations. If this style favors centralized management, a company-wide PSM system implementation strategy may be highly effective.

A company-wide approach has key benefits, including:

- ensuring consistent application of PSM systems throughout your company and
- minimizing resource requirements for program development and some activities through economies of scale.

In addition, a company-wide PSM system is likely to yield ongoing programs that can be supported by a relatively small, centralized group of experts.

At the same time, companies whose PSM system requirements vary widely among locations may not find a centralized, company-wide approach practical, unless PSM system performance is uniformly poor throughout the organization. In any case, some central coordination will help identify common issues and facilitate the sharing of knowledge and experience among locations.

Depending on the findings of your PSM system assessments, you may wish to consider a company-wide (or centralized) approach if either of these situations exists:

- **The current performance of the PSM system is relatively low.** If this is the case, it probably means that the knowledge and expertise are not sufficient for local implementation to be effective. A more practical approach involves a small centralized team and a plan that emphasizes training of local staff to upgrade necessary skills.

- **Manufacturing processes are similar in each location.** If your company's processes are essentially comparable throughout the organization, you may have a good opportunity to minimize process safety resource requirements. For example, a detailed program at one facility can easily be copied and installed at other facilities without significant modification. If this approach runs counter to corporate culture (e.g., each location is used to developing its own programs with independence), you will need to build in a degree of local implementation to ensure local buy-in.

A centralized approach generally calls for establishing multiple teams under the direction of a single implementation manager or champion to ensure coordination. The overall structure of a company-wide task force and its organization will depend on your company's management style and the implementation strategy you adopt. Some companies routinely use independent project teams for any significant task, while others establish teams within existing departments.

The centralized team(s) will approach process safety element by element. The teams develop detailed guidelines, procedures, and standards for each element, which can then be installed by facility management. However, in handling one element at a time, teams must be sure to incorporate a consistent approach to those elements that are interrelated (e.g., training) and to the management system characteristics (e.g., documentation) that apply to the whole PSM system.

In considering this method, keep in mind that these teams must include staff members who are familiar with all the operations of the company to ensure consistency. In addition, team members as a rule should have experience in operations, design, maintenance, and safety management.

Note: Consider the practicality of assembling and managing people with these kinds of qualifications in your company; this may help you determine whether a company-wide approach is feasible. Single divisions may offer stronger possibilities, since they usually have more common manufacturing processes and equipment.

Facility-specific approach

A facility-specific implementation strategy relies on local expertise and calls for a number of local process safety teams to work in parallel. One benefit of this approach is that it can provide for very rapid implementation of PSM systems. In addition, facility-specific programs can be adapted to local requirements such as different management styles, equipment, or local regulation. The downside is that this may result in programs that are very different in each location, reducing the possibilities of learning from one another.

Facility-specific approaches to implementing PSM systems make sense if you have identified significant variations among locations in one or more of the following areas:

- current performance of PSM systems,
- potential process hazards,
- local regulatory requirements,
- manufacturing processes,
- process technology and equipment, or
- local management styles.

Note: As previously noted, facility-specific approaches tend not to succeed where the overall current performance of PSM systems is poor. Local staff will not have the necessary knowledge of or experience with PSM systems to develop and implement a program without considerable outside assistance. Even if you have identified significant variations, if your team decides that overall PSM system performance is low, you should consider a company-wide strategy rather than a facility-specific approach.

Facility-specific implementation requires formation of a local team at each of your company's facilities. As a general rule, these teams are composed of resident staff and report to the facility manager. A typical local team would include the facility safety manager and representatives from operations, maintenance, and engineering.

More often than not, teams such as these benefit from outside assistance, either from other company experts or from consultants or engineers who can supplement their expertise. In any case, these teams, either directly or through facility managers, should report to a company implementation manager or champion to monitor progress, ensure consistency, and facilitate cross-fertilization of efforts through exchange of experience.

Note: Where individual facilities have strong programs in specific aspects of PSM systems, a facility-specific approach may involve identifying "best practice" programs and facilitating cross-communication among locations.

Hybrid approach

Neither the company-wide nor the facility-specific implementation model offers a magic solution. In fact, many companies incorporate elements of both strategies into a hybrid approach that better meets their individual needs. A hybrid approach addresses some PSM elements centrally and others locally, reflecting current PSM performance. In many real-world situations, this offers the best way to take advantage of what already exists. Where the performance of different elements varies (e.g., some may be poor throughout the company, while others are either strong throughout the company or at particular locations), a hybrid approach may be the only effective means of addressing gaps systematically.

Hybrid approaches are generally more difficult to manage because the relationships and division of responsibilities among the PSM project teams and other management varies from element to element. This can lead to confusion within both the PSM teams and among other managers, which can result in misunderstandings and wasted work.

For a hybrid program to succeed, the management of the teams must be thoroughly planned in advance. The result should be a network of corporate and local teams that work with well-defined responsibilities. Overall coordination generally lies with the corporate implementation manager, who will manage a combination of staff members working on specific elements and staff members working on particular facilities (see Table 4.19). Local facility management represents a third dimension, and responsibility for coordination at the local level will generally be with a local implementation manager. Hybrid programs require carefully designed communication programs specifically designed to keep all the interested parties up to date on all activities.

Note: For any of the three general approaches discussed above, a "phased approach" could also be used in certain circumstances. In some cases companies have found that certain requirements might initially rely heavily on other actions being implemented first. In such cases a minimum requirement(s) may be established during Phase 1 of the implementation, with an increased (best in class) requirement being adopted in Phase 2 (without even having identified at this stage when Phase 2 is going to happen).

Table 4.19 Example of Responsibility Matrix for Hybrid Approach

Team	Responsibility	Leader
Corporate MOC team	Develop MOC system for use at all sites	Facility Maintenance Supervisor
Corporate training team	Develop training system for use at all sites	Corporate Human Resources Manager
Facility X PSM team	Adapt Facility X PHA system for use at all sites	Facility Safety Manager
Team at each facility	Adapt own preventive maintenance system to meet PSM needs	Facility PSM Coordinators

Define priorities

No company today has the luxury of having unlimited resources to implement PSM programs, and few have the capacity to tackle everything at once. Clearly, priorities must be set for each of the actions required by the system to ensure orderly and rational implementation.

Having assessed the current status, prepared the organization for change, and designed a new system, the organization should be well equipped to establish workable criteria for setting priorities. Such criteria might include:

- compliance with regulations,
- the extent of potential hazards,
- relevance to facility conditions,
- compliance with corporate policies,
- integration with business processes (see Chapter 7), and
- the need for new electronic tools (e.g., databases, computerized maintenance management system).

Before determining which of these (or other) criteria make sense for your company's situation, it may be useful to consider some of the reasons why each may be important.

Noncompliance with regulations exposes your company to the risk of financial penalties arising from fines, restrictions, and/or closure of operations. In addition, the company can suffer serious adverse publicity as a result of regulatory violations, which in tum can affect customers, shareholders, and employee goodwill. For example, you may have an MOC program that is lacking an aspect required by OSHA, such as the time period of the change. Resolving a noncompliance situation may increase the priority for implementing a particular element at a specific facility.

The extent of potential hazards varies among facilities and among units within a facility. If two facilities have similar inadequacies in their PSM program, but one has a greater hazard potential, the benefit of improvement will be greater at the higher hazard facility. For example, a strong MOC procedure will be more important in the processing of reactive chemicals than in the processing of aqueous latex products. In the absence of other considerations, PSM priorities can be set to reflect the level of hazard at each unit.

Relevance of each element to facility conditions will vary. Some actions will produce a greater improvement in the level of process safety than others. Priorities for action items could be set according to the improvement in safety achieved. For example, addressing MOC may have greater impact than addressing facility siting at an existing facility.

Corporate policies governing safety management must be reflected in a company's PSM program.

Note: In setting priorities for action, you must also consider available resources. You may identify an action step that produces the single largest improvement in process safety, only to discover that it requires all of your available resources. Alternatively, you may be able to spread your efforts among a number of lower priority actions that together have a greater impact on overall safety performance.

Note: Remember as you set priorities that you and your team must consider the expected benefit to the company, division, or unit as a whole, rather than simply addressing a single gap. Obviously, any specific highly ranked risk identified in the course of the assessment should be addressed as determined by the company and/or facility risk management program guidelines; however, be careful not to let anomalous findings skew your perspective on broader priorities. For example, the absence of a capital project review process is a significant gap. However, if there is only one project per year, this gap may have lower priority than an existing, but seriously deficient, operator training program.

Criteria such as those suggested above can be usefully applied to the implementation strategy you have selected. In assessing your current PSM status, you will have gained considerable insight into which facilities warrant priority attention; similarly, the assessment process sheds light on those PSM elements whose improvement will yield the most productive results.

Whether you have chosen a facility-specific, company-wide, or hybrid approach, it may be helpful to consider your priorities in terms of both facilities and PSM elements. The goal here is to help determine what needs doing, in what order, and with what level of effort.

Prioritize facilities

In the course of assessing your company's current PSM status, you and your team have almost certainly gained a clear sense of which facilities pose the greatest risk, whether by virtue of inherent process hazards, human factors, management systems, or a combination. As you set priorities for implementation, you should closely review information gleaned from the assessment tasks. In addition, you should try to validate or flesh out your impressions through quantitative analyses that can help identify priority facilities.

Simply stated, the extent of potential hazard depends on how much hazardous material a facility uses and the conditions under which it is contained. The quantity of material stored provides a measure of the size of any release that could result from a loss of facility integrity. The conditions under which it is contained can influence the extent of any damage that might result from the release. Appendix V provides an example of an initial facility ranking process.

However, total quantities of hazardous materials do not, on their own, provide an entirely reliable measure of potential hazard. It is more useful to consider quantities of material within sections of the facility that can be isolated. The amount of material within these individual facility sections usually represents the largest credible release that could occur. Some examples of facility sections that may be isolated include tank farms, unloading racks, and separate process buildings.

You can quickly identify these facility sections by reviewing process flow diagrams and valving arrangements. Isolation points may be defined by control valves or powered block valves that can be remotely activated. PHA techniques can help you identify the maximum credible accident scenarios. (Note that manual valves should not be considered reliable isolation points unless they are located to be accessible following a major accident. However, remotely activated valves can only be considered reliable isolation points if there are adequate reliability engineering and maintenance programs in place.) Alternately or in addition, you can consider using (1) EPA's "Hazard Assessment" requirements (per 40 CFR Part 68, Subpart B) and/or *Risk Management Program Guidance for Offsite Consequence Analysis* to identify and quantify credible scenarios and (2) the Dow Fire and Explosion Index method to quantify scenarios (see *Dow's Fire & Explosion Index Hazard Classification Guide, 7th Edition,* for more information).

You should be able to estimate the quantities of material contained within a section from mechanical and operating data. You should also consider operating conditions, which should be available from the facility mass balance or from actual operating data. Simple hazard models can predict the size of vapor clouds, radiation hazards from fires, and explosion overpressures. Such models are available from a number of sources.

Note that facility siting issues related to fires, explosions, and toxic releases and their potential impacts on building occupants are addressed in detail in (1) *Guidelines for Evaluating Process Plant Buildings for External Explosions, Fires, and Toxic Releases, 2nd Edition*, (2) API RP 752 (*Management of Hazards*

Associated with Location of Process Plant Permanent Buildings), and (3) API RP 753 (*Management of Hazards Associated with Location of Process Plant Portable Buildings*).

Once you have ranked your company's facilities in terms of inherent hazards, compare these results against the gaps you have identified in the PSM assessment tasks. By looking at both PSM element maturity and facility hazards, you gain a more complete picture for prioritizing facilities (see Table 4.20 for an example of a graphical approach).

For example, a high-hazard facility that has a relatively poor overall PSM performance history (e.g., Facility X in Table 4.20) should command top priority; low-hazard facilities with relatively sound PSM programs (e.g., Facility Y) fall at the other end of the spectrum.

In between, there will almost certainly be a number of cases that are far less clear cut. For example, you may find situations in which PSM programs are well developed but fail to fully address a high-hazard process or a worst-case scenario; in others, a facility engaged in relatively low-hazard operations may have a proportionately low level of PSM systems.

Table 4.20 Example Analysis of Selected PSM Gaps and Priorities

PSM Element	PSM Element Maturity		
	Embryonic	**Developing**	**Mature**
Management of change	X	*Y* Z	
Training	Z	X	*Y*
Process knowledge management		Z X	*Y*
Compliance with standards		*Y*	Z X
Conduct of operations	*Y* X Z		

Z = Facility Z
Y = Facility Y
X = Facility X

Hazard code: Large letter size indicates greater hazard. For example, high hazard materials (X) include phosgene, carbon disulfide, ethylene oxide, etc. Moderate hazard materials (Z) include toluene, sulfuric acid, etc. Low hazard materials *(Y)* include soda ash, alum, etc.

Note: There is no substitute for judgment and experience. Consider these quantitative exercises as tools or methods to guide the team's determination of the right priorities, not as absolute formulas. In addition, keep in mind that the goal is to improve the overall status or performance of process safety as part of a continuing effort, not as a single action, event, or "silver bullet."

Prioritize elements

To establish priorities among the elements of the PSM program, you will need to consider two factors: (1) the level of noncompliance and (2) interrelationships with other elements.

Levels of noncompliance

Consider your company and the findings of your PSM assessment for each of the process safety elements in terms of these levels of compliance:

- A formal system exists and is operational
- An informal system is in place
- A system exists but is not followed
- An incomplete system is in place but needs upgrading
- No system exists

Each of the noncompliance situations is fundamentally different, and bringing each of them into compliance will require a different approach and series of actions:

- *Where an effective informal system exists and is followed, the issue is one of style, not substance.* A facility or unit may have a strong safety culture and sound safety practices, but its managers lack the habit of formal documentation, or simply don't think it is important. Assuming that safety performance meets applicable standards, you will probably assign cases like these a relatively low priority, compared with other noncompliance situations. Cases like these are also often the easiest to fix; since the fundamentals are already in place, what's required is to formalize the informal system by preparing and implementing documentation procedures. On the other hand, care should be taken not to underestimate the effort or skip vital communication steps with stakeholders. Effective communication, participation, ownership, and buy-in will be keys to effective implementation. Poor implementation could result in worse PSM performance than before.
- *Where a system is not followed or does not exist, the effect on safety is the same and should command the same priority for attention.* Keep in mind, however, that the underlying causes for these situations may be very different. In any case, when you identify these noncompliance situations, you should carefully consider *why* they exist; this can yield valuable

qualitative information about potential obstacles to effective implementation.

- *If an element has a formal system that isn't followed*, this suggests either a problem with the system itself (e.g., flawed design, impractical, insufficient detail or explanation) or a failure of management (e.g., inadequate supervision, lack of training, no follow through or enforcement, inadequate resources). Whatever the cause, you will want to make note of it and try to address it in your implementation plan.
- *If an element has no system at all*, it may indicate that the facility manager does not fully understand what the element requires, or is for some reason unable to carry out his/her responsibilities. In addition, a weakness in corporate oversight, direction, or resource allocation may contribute to the problem. Either way, the plan you develop must take these gaps into account.

Interrelated PSM elements

Some PSM elements are relatively easy to isolate, while others are integrally related, touching virtually every aspect of the program as a whole. In any program you should devote the highest priority to those elements that impact other parts of the program. This will yield greater efficiency for your PSM effort and improved consistency for the resulting system. Two elements of the RBPS PSM system model with particularly widespread impact are process knowledge management and hazard identification and risk analysis. Other elements with significant cross-element impact include training, MOC, and operating procedures.

In addition to considering those elements that interrelate, you should pay particular attention to those that affect your operations most directly. For example, if your company has a major expansion program underway, you would assign the MOC (particularly for capital projects) and the compliance with standards elements a higher priority than if your company has deferred all major capital projects.

Limitations on the scope of the plan

Defining priorities helps determine the limitations of your company's plan, which in turn makes it more manageable. Remember that your goal at this stage is to develop a workable prototype, not an all-inclusive, perfect blueprint. If your plan is properly thought out and soundly put together, it should be adaptable to a wide range of contingencies – not all of which need to be explicitly addressed.

At a minimum, your plan should address those activities that are covered by the OSHA PSM and/or EPA RMP regulations, and by equivalent local regulations and corporate standards. This means that all manufacturing or storage operations handling specified materials in excess of threshold quantities must be included in the program. Once your plan has been developed, tested, and refined, you may wish to consider expanding it to include activities that may not be governed by regulation but that would benefit from PSM principles and practices. For example, the current OSHA and EPA regulations cover only certain operations that involve one or more listed chemicals. However, a comparable operation handling an unlisted material might be considered for comparable treatment because even if a chemical is not on the OSHA or EPA list of highly hazardous chemicals, that does not mean it is not hazardous to employees, the environment, and/or neighboring communities.

In the short term, however, your plan may be most effective if you focus on the specific needs and priorities you have identified in the course of your work. If you clearly define the scope of your plan and it directly addresses specific needs, your efforts will be far more successful than if you try to do all things for all people. Moreover, focusing on priority needs will almost certainly provide a sound basis for expansion or adaptation; for example, procedures developed to address training for operators using high-hazard materials at one facility (see Facility X vs. Facility Z in Table 4.20) should be readily adaptable to lower-hazard substances at another facility – far more so than the other way around.

Develop, communicate, and gain approval of the implementation plan

Just as you probably presented a preliminary plan at the outset of the PSM initiative, you may want to think in terms of a formal presentation to your company's management when the detailed plan is complete. Remember that senior management's buy-in will be essential to successful implementation, as will the endorsement of line managers at the operating level who will be directly affected by your plan.

The way you present the plan depends a great deal on your company's style, your management's preferences, and the expectations you have established early in the process. In a hierarchical organization, senior management may expect to see and approve the plan before it is more widely disseminated. Other executives prefer to review materials after line management has seen and commented on them.

Either way, both senior management and line personnel should have the opportunity to review the PSM plan and discuss it with the team. As a general rule, the more input and commentary you can incorporate into the finished product, the better its chances of approval – and of successful implementation – since it will reflect the interests of the people affected by it.

Note: If you decide on a formal presentation, consider having the full PSM team participate. Similarly, written communications concerning the plan (if, for example, it is broadly distributed for comment) should originate with the team. This not only acknowledges their contribution; it also tangibly demonstrates the interdisciplinary nature of the challenge you have undertaken.

A formal presentation of the detailed plan might include the following topics:

1. Brief review of the preliminary plan (presented previously)
2. Discussion of the process and tasks to date
3. Overview of the PSM assessment findings
4. Major variations from the initial proposal
5. Discussion of the implementation strategy selected and the rationale behind it
6. Discussion of priorities, by PSM element and by facility
7. Summary of the plan
   a. tasks required
   b. schedule
   c. resource requirements
   d. accountability and reporting structure
   e. expected results
8. Summary of plan benefits
9. Summary of support requested
10. Questions and answers

An example of Item 7 (summary of the plan) is provided in Appendix VI.

If you have circulated drafts of the plan, or portions of it, to your company's decision-makers, a full-blown presentation may not be necessary. However, you may find the discipline of following an outline such as the one above useful in organizing less formal discussions, either individually or with groups.

It's important to remember to "close the sale." Your goal for this discussion is not only to win approval, but also to gain a commitment for the resources you will need to move the initiative forward. To achieve this goal, you should devote particular attention to summarizing the benefits of the plan and the kinds of support you are requesting; these two points, taken together, create the essence of your sales proposition: Now you know what you will get and what it will cost you.

To help define these benefits as crisply as possible, one possible approach is to ask each team member to answer this question: "Why is this plan the best possible way for our company to implement PSM?" Then ask each one to list every positive characteristic the plan offers and every positive effect the company can expect from it. You will almost certainly find that the results of this brief exercise can be easily grouped into categories, each of which probably represents a selling benefit.

Then (1) consider these selling benefits within the context of your company's business priorities, (2) match them against the goals management has established for PSM, and (3) try to distill the comments into a few key points that summarize the long-term value of your plan to the company as a whole.

4.5.4 Confirm the Tools and Associated Resources

At this point it will be time to confirm that all of your PSM system tools are ready for the rollout, along with the associated resources required. To this end, this step should include:

- preparing the final tools (hard copy or electronic), forms, and supporting materials (e.g., intranet/network structure, spreadsheets, files) needed to execute various PSM system activities;
- if you are using any electronic workflow, documentation, or corrective action systems for PSM activities, confirming that fully tested systems with sufficient network resources dedicated to their operation and maintenance have been established; and
- ensuring resource commitments from infrastructure departments (e.g., information technology for electronic systems, human resources for their training system responsibilities, purchasing for their contractor and MI responsibilities).

4.5.5 Complete the PSM Procedures

Prior to the PSM system rollout, the written programs/procedures (see Section 4.4) need to be fully refined and, in many cases, pilot tested (see Section 4.5.2). Ensure that they are ready to go; if not, adjust the implementation plan schedule accordingly.

4.5.6 Provide PSM Rollout Training

Most long-time PSM practitioners have personally experienced multiple cases where a new, well-conceived, well-written PSM element system was rolled out without adequate training, and either had to be pulled back and reworked or restarted – or failed. It is vital that (1) facility personnel are aware of new systems and their requirements; (2) personnel directly involved in the system fully understand the new system, its requirements, and their roles/responsibilities; and (3) there are an adequate number of subject matter experts to support the day-to-day operation of the system, as well as manage the evolution and any troubleshooting. To accomplish these objectives, PSM system rollout training should include the following:

1. **Develop a plan for PSM rollout training.** Consider:
   a. what overall system training is needed for facility personnel,
   b. what PSM elements warrant rollout training,

   c. what type of training will be adequate and effective (e.g., classroom, computer-based, or a combination),
   d. what groups and/or specific personnel need to be trained in the overall system and individual elements,
   e. what level of training (e.g., awareness, detailed, or expert) is required for each group of trainees, and
   f. when the various training courses should be delivered.
2. **Develop PSM rollout training materials.** Determine what training materials are needed based on the training plan, assign responsibilities and timing for developing them, and ensure that adequate review (perhaps including pilot testing) is provided for the materials.
3. **Deliver PSM rollout training.** Initially, company-wide awareness sessions can be conducted (in person or by video conference) and/or bulletins issued to advertise the PSM implementation team, goals, and effort as a means to complement the formal training program. Then, provide the PSM rollout training courses/sessions based on the training plan and schedule. Be sure to obtain feedback from trainees during the training and closely review that input to determine whether revisions to any elements or the system, or adjustments to the implementation plan, are warranted.

Note: Depending on priorities and risk, some new PSM elements may need to be rolled out and implemented on a case-by-case basis as soon as it is possible to do so.

Note: Chapter 14 of the RBPS book is an excellent resource for planning, preparing, and delivering effective training. Consider reviewing and utilizing the guidance it provides to ensure that your PSM rollout training is effective.

4.6 MONITOR THE PSM SYSTEM'S IMPLEMENTATION, INITIAL PERFORMANCE, AND PROGRESS

Just as it is important to assess process safety performance or status before implementing a new PSM system, it is important to monitor the initial implementation and performance of the new system. So, in addition to establishing ongoing measurement, management review, and auditing requirements for the new elements/system, consider:

- **Establishing a plan/means for collecting PSM metrics and other feedback and data on the performance of the new elements and system.** During initial implementation, more frequent collection of PSM metrics may be appropriate, and close attention should be paid to ensure that the metrics collection is correct and effective. In addition, there may be additional data that can be collected on the element/system performance during this period to aid in evaluation of performance. Finally, it will be help ensure success to proactively reach out to system practitioners frequently to determine whether there are issues that need to be addressed or ideas for improvement that should be considered (either on a priority basis or as future enhancements).

- **Developing a plan for the initial PSM management reviews.** As previously discussed in this chapter, management reviews are vital to continuous improvement of the new PSM system. However, since they are likely to be a new activity for your facility and the management team, extra attention should be paid to ensure that the initial management reviews are successful and add value. Such extra attention may include the following:
    - Assistance from the facility PSM team or coordinator in preparing and reviewing presentation materials for the management review meetings
    - Dry runs of presentation delivery, with the facility PSM team or coordinator posing possible management questions and comments
    - Participation by the facility PSM team or coordinator in the initial management review meetings
    - Ensuring that the initial management review meeting minutes and recommendations set the tone and provide a good model for the subsequent management reviews
- **Creating a plan for the initial audit of the PSM elements and system.** In addition to ongoing, periodic audits as previously discussed in this chapter, it will likely be beneficial to perform early, additional audits of some or all of the new PSM elements and the overall system. To that end, consider what audits would be beneficial, schedule the audits, perform them, and implement the audit recommendations to improve the PSM elements and system.

If you either elected to phase in certain elements initially and others later or subsequently decide to add new elements to the system, see Chapter 5 for guidance on this subject.

If your initial or ongoing monitoring of the performance of the PSM elements or system detects performance gaps, see Chapter 6 for guidance on this subject.

Note: In addition to the guidance on implementing a new PSM system provided in this section, see Chapter 23 of the RBPS book for a specific example of using the RBPS approach to develop and implement a new PSM system.

4.7 REFERENCES

4.1 Center for Chemical Process Safety of the American Institute of Chemical Engineers, *Chemical Process Safety Management – A Challenge to Commitment*, New York, New York, 1988.

4.2 Center for Chemical Process Safety of the American Institute of Chemical Engineers, "Guidelines for Technical Management of Chemical Process Safety," *Plant/Operations Progress*, New York, New York, Vol. 10, Issue 2, April 1991, pp. 65-68.

4.3 Center for Chemical Process Safety of the American Institute of Chemical Engineers, *Guidelines for Risk Based Process Safety*, John Wiley & Sons, Inc., Hoboken, New Jersey, 2007.

4.4 Energy Institute, "High Level Framework for Process Safety Management," London, England, UK, 2010, www.energyinst.org/technical/PSM/PSM-framework.

4.5 Center for Chemical Process Safety of the American Institute of Chemical Engineers, *Essential Practices for Managing Chemical Reactivity Hazards*, John Wiley & Sons, Inc., Hoboken, New Jersey, 2003.

4.6 Center for Chemical Process Safety of the American Institute of Chemical Engineers, *Guidelines for the Management of Change for Process Safety*, John Wiley & Sons, Inc., Hoboken, New Jersey, 2008.

4.7 Center for Chemical Process Safety of the American Institute of Chemical Engineers, *Guidelines for Process Safety Metrics*, John Wiley & Sons, Inc., Hoboken, New Jersey, 2009.

4.8 ANSI/API Recommended Practice 754, *Process Safety Performance Indicators for the Refining and Petrochemical Industries*, American Petroleum Institute, Washington D.C., April 2010, www.publications.api.org/.

4.9 Center for Chemical Process Safety of the American Institute of Chemical Engineers, *Guidelines for Auditing Process Safety Management Systems, 2nd Edition*, John Wiley & Sons, Inc., Hoboken, New Jersey, 2011.

5

INTEGRATING NEW ELEMENTS INTO AN EXISTING PSM SYSTEM

5.1 DEVELOPING A NEW ELEMENT

When a site/company elects to phase in certain elements initially and others later, or subsequently decides to add new elements to the existing PSM system, the basic steps discussed in Chapter 4 apply:

1. **Develop the design specification for the new element.** Keep in mind that the design specification for each PSM element will depend on the level of detail required for each element. See Section 4.1 for more information.
2. **Create element workflows (as appropriate).** If the steps involved in implementing the element activities, or if the interactions among the steps and the workgroups involved in the implementation activities are quite complex, developing a workflow is beneficial. See Section 4.2 for more information.
3. **Estimate the element workloads and necessary resources.** Without an adequate understanding of the workloads and resources required by the new elements and their impact on the overall PSM system, the new elements are likely to fail. See Section 4.3 for more information.
4. **Develop the written programs and procedures for the element.** A PSM system and its individual elements must be described in written programs and procedures to be sustainable. See Section 4.4 for more information.
5. **Roll out the element.** The steps culminating in the rollout of a new PSM element may include (a) gathering implementation input from all stakeholders; (b) pilot testing the element in phases (depending on its size and complexity); (c) developing an implementation plan with a defined scope and approach; (d) confirming the need for and availability of any special tools/resources; (e) confirming that the written procedures are refined, tested, and ready; and (f) developing and delivering training on the new element to personnel who need to be aware of or involved in its application. See Section 4.5 for more information.

6. **Monitor implementation and initial performance.** Additional monitoring of the initial implementation and performance of the new element may be appropriate. See Sections 4.6 and 5.4 for more information.

Finally, keep in mind that the effort and resources required for accomplishing each of these steps will obviously be less, and may be much less than for implementing a whole new system. In some cases, it may even be decided that one or more of the above steps is not adding value due to the scope of the effort/resources involved.

5.2 INTEGRATING NEW ELEMENT ACTIVITIES INTO EXISTING ELEMENTS

Integrating new element activities into existing PSM elements is a more evolutionary change that may be needed due to factors such as organizational structure changes, poor performance of an element (based on monitoring), and suggestions for improvement. In addition, determining that some of the "expanded" elements and associated activities in the RBPS model would be good additions may be a driver for such changes.

Once again, the steps listed above apply, but the effort and resources required will generally be less than for implementing a new system, and not all the steps will be necessary. In addition, if the organization's safety culture and the element implementation are strong, even less effort and resources should be required.

It is important, however, not just to continue "adding to the load" of the organization unless the anticipated results are worth the additional effort. Therefore, adding new element activities should only be done after (1) evaluating the cost/benefit of the activity and (2) considering whether there are any existing activities that are no longer needed or will be replaced, and therefore can be stopped.

5.3 IMPLEMENTING NEW RBPS ELEMENTS

The implementation of new RBPS elements into an existing PSM system represents a unique challenge and opportunity for an organization. For example, if the existing system is focused on regulatory compliance, these can represent a large expansion of the system. In addition, the new RBPS elements listed below (i.e., elements that are not in the OSHA/EPA regulatory model or the 1995 CCPS model [discussed in Chapter 4 of this book]) primarily deal with "soft," people-related areas rather than "hard," process-related or technical areas, and therefore are likely to require different implementation approaches:

- Process safety culture (Chapter 3 in the RBPS book)
- Compliance with standards (Chapter 4 in the RBPS book)

- Process safety competency (Chapter 5 in the RBPS book)
- Stakeholder outreach (Chapter 7 in the RBPS book)
- Conduct of operations (Chapter 17 in the RBPS book)
- Measurement and metrics (Chapter 20 in the RBPS book)
- Management review and continuous improvement (Chapter 22 in the RBPS book)

The RBPS book suggests an implementation approach for RBPS elements where the degree of rigor designed into each work activity is tailored to risk, tempered by resource considerations, and tuned to the facility's culture. Thus, the degree of rigor that should be applied to a particular work activity will vary for each facility, and likely will vary between units or process areas at a facility. Therefore, to implement new RBPS elements, the following steps are recommended:

1. Assess the risks at the facility, investigate the balance between the resource load for RBPS activities and available resources, and examine the facility's culture. See Section 2.2 of the RBPS book for more information on this step.
2. Estimate the potential benefits that may be achieved by addressing each of the key principles for each RBPS element to be implemented.
3. Based on the results from steps 1 and 2, decide which essential features described in the RBPS book are necessary to properly manage risk.
4. For each essential feature that will be implemented, determine how it will be implemented and select the corresponding work activities described in the corresponding chapter of the RBPS book. Note that this list of work activities cannot be comprehensive for all industries; readers will likely need to add work activities or modify some of the work activities listed in the chapters.
5. For each work activity that will be implemented, determine the level of rigor that will be required. Each work activity listed in the RBPS book chapters is followed by two to five implementation options that describe an increasing degree of rigor.
6. Apply the six steps summarized in Section 5.1, as appropriate, to design, develop, roll out, and monitor implementation of the new element and the associated work activities.

Guidance on implementing each of these new RBPS elements and some implementation examples are provided in the following sections (some of the information is borrowed from the RBPS book). Detailed discussions of each element are provided in the corresponding RBPS book chapters, which provide an element overview, key principles and essential features, possible work activities, examples of ways to improve effectiveness, element metrics, management review, and references.

5.3.1 Process Safety Culture (Leadership, Commitment, and Accountability)

Process safety culture has been defined as "the combination of group values and behaviors that determine the manner in which process safety is managed" (Ref. 5.1). More succinct definitions include "how we do things around here," "what we expect here," and "how we behave when no one is watching."

A culture develops as a group identifies certain attitudes and behaviors that provide common benefit to its members; in this case, attitudes and behaviors that support the goal of safer process operations. As the group reinforces such attitudes and behaviors and becomes accustomed to their benefits, these attitudes and behaviors become integrated into the group's value system (Ref. 5.2). In an especially sound culture, deeply held values are reflected in the group's actions, and newcomers are expected to endorse these values in order to remain part of the group.

The process safety culture of an organization is a significant determinant of how it will approach process risk control issues, and PSM system failures can often be linked to cultural deficiencies.

An organization's process safety culture is founded on its underlying values regarding process safety. Successful cultural change requires that (1) expectations of new attitudes and behaviors be communicated and reinforced, (2) these new attitudes and behaviors demonstrate successful results, and (3) the members of the organization recognize and appreciate the resulting successes (Ref. 5.2).

Leadership's role is vital, as it is primarily a management responsibility to (1) set the standards (for both process safety and individual behaviors), (2) set the tone regarding the importance of process safety and a sound culture, (3) provide the resources required to meet expectations, and (4) provide continuous and positive reinforcement. In short, leaders can enable and nurture a sound safety culture, but cannot mandate it. As Dr. Edgar Schein said: ". . . one could argue that the only thing of real importance that leaders do is to create and manage culture . . ."

Acceptable behaviors must be modeled at all levels of the organization through leadership by example. The rationale for, and anticipated benefits of, expected behaviors must be made evident to all. Positive reinforcement and accountabilities for expected behaviors must be clear and certain. By consistently reinforcing positive behaviors and linking them to the important benefits they bring, management should be able to gradually shift the values of the organization in a positive direction, advancing the organization from a rule-driven culture to a value-based culture.

Many organizations have successfully established sound process safety cultures. Often, these cultures have been developed in response to, and are reinforced by frequent reference to, significant loss events in the company's past.

Other organizations may take justifiable pride in an exemplary process safety record and may seek to inspire employees to maintain their diligence and efforts in order to preserve that record (while seeking to avoid the complacency that past successes may inspire). In either case, commitment by each member of the organization is vital to achieving and sustaining a sound process safety culture.

Implementation of a new process safety culture element can involve many steps and *will* require several years to fully implement in a moderate to large organization. Some of the steps to consider include the following:

Collect culture evaluation evidence

1. Conduct a broad, confidential survey of employees at all levels to assess perceived cultural strengths and weaknesses (for example, with questions based on the 12 essential features of a sound process safety culture [see Section 3.2 of the RBPS book]).
2. Conduct follow-up interviews with a cross-section of the organization to identify possible underlying causes of the survey results.
3. Perform work observations to assess how well employees at all levels and in all types of work "follow the rules."
4. Assess the organization's performance based on PSM and HSE leading indicators.

Assess HSE technical performance

5. Review information sources such as incidents and incident investigation results, audits and assessments, and action item completion history that relate to both HSE technical performance and process safety culture.

Identify and address cultural strengths/weaknesses

6. Assemble, collate, and analyze all of the above information to:
    a. identify cultural strengths/weaknesses (e.g., as compared to the 12 RBPS essential features of a sound process safety culture) and
    b. identify contributing causal factors for the identified strengths/weaknesses.
7. Determine ways to address the identified cultural weakness, as well as to maintain or build on the strengths.
8. Roll out the program throughout the organization.
9. Monitor results and follow through to continuously improve.
10. Repeat steps 1 through 9 every 2 to 3 years, until a sound process safety culture is established and sustainable.

Cultural change

One way to gain insight into your organization is to assess how it "operationalizes" safety. In other words, is it reactive, dependent, independent, or interdependent in terms of its attitude toward safety? The DuPont Bradley Curve shown in Figure 5.1 graphically illustrates this concept. It provides additional detail on each of these four attitudes and how they should relate to safety performance and safety culture.

Another culture model, which introduces the Energy Institute's "Winning Hearts and Minds" program (Ref. 5.3), is shown in Figure 5.2.

Readers may also find the following excerpt from "Take the FUN Out of Process Safety" (Ref. 5.4) helpful when considering how to implement cultural change in their organization:

"At the core of Daniels' approach [to behavioral safety] is the ABC Model, where A stands for antecedents, B for behaviors, and C for consequences. Antecedents create favorable circumstances for a particular behavior to occur once. Consequences are the outcome of a behavior. Daniels suggests that consequences, perceived or previously experienced, play a significant role in managing future behavior (i.e., shaping decision-making behavior).

"Consequences can be characterized by three dimensions:

- timing – will the consequences occur immediately (I) or at some time in the future (F)?
- probability – is the probability of the consequence occurring certain (C) or uncertain (U)?
- type – is the consequence positive (P) or negative (N)?

"Thus, consequences can be Immediate, Certain, and Positive (ICP), or Future, Uncertain, and Negative (FUN).

"The vital cultural change requires that the FUN be taken out of process safety. The consequence of taking the FUN out would be that process safety incidents are prevented early, and willingly, because the managers making the resource decisions clearly see the immediate, certain, and positive benefits of investing in projects that improve process safety. The consequences of leaving FUN in is that the skeptical decision-makers may choose to not make an investment because they cannot see its benefit."

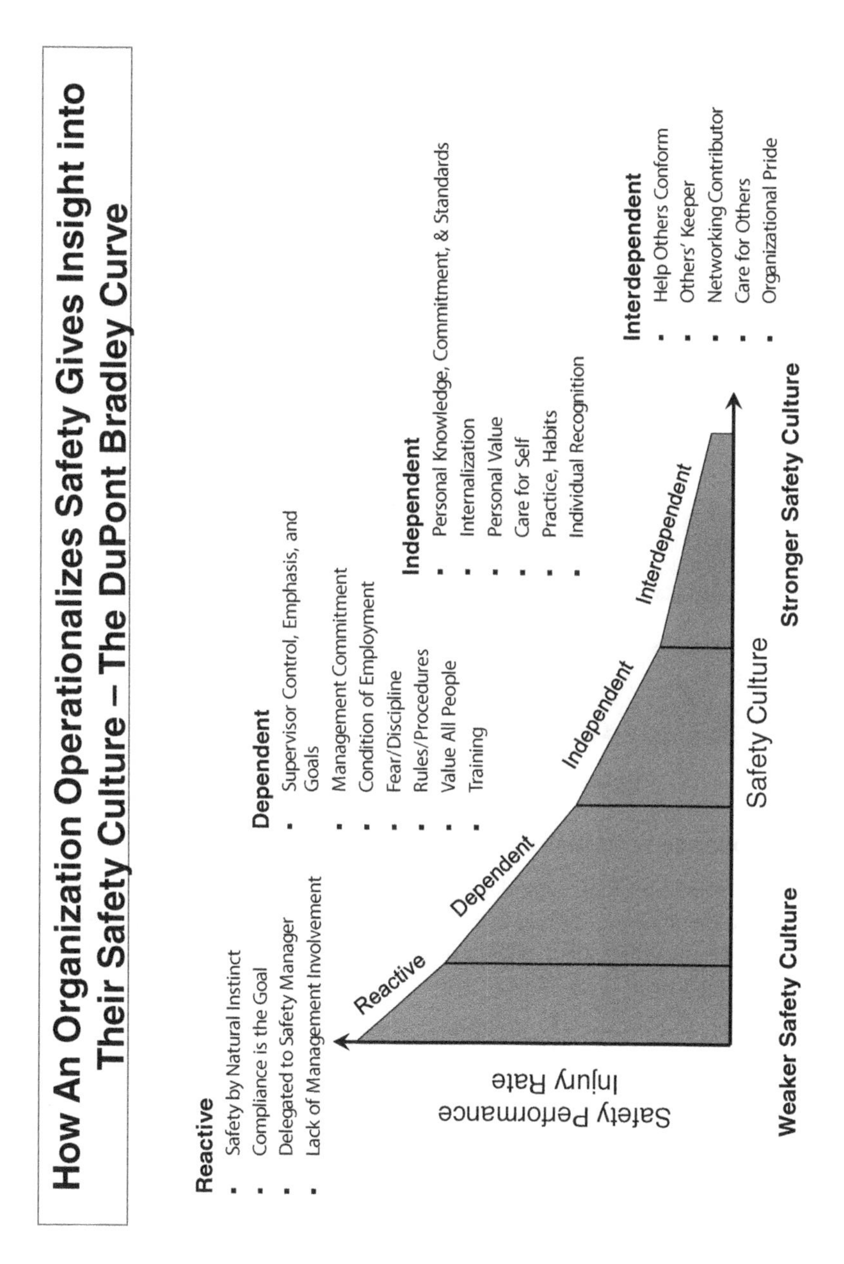

Figure 5.1 The DuPont Bradley Curve

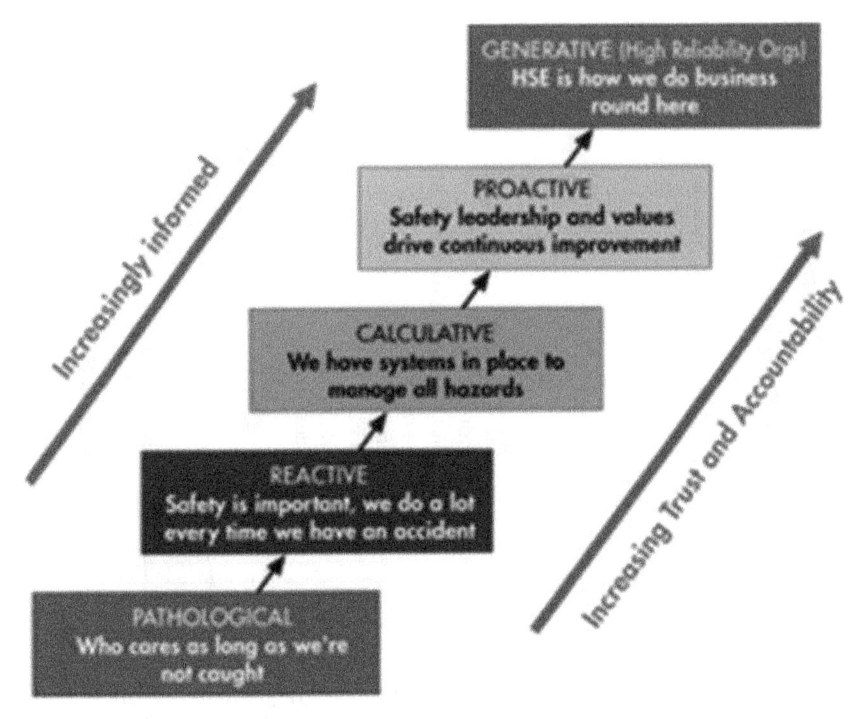

Figure 5.2 The HSE Culture Ladder

5.3.2 Compliance with Standards

The RBPS Compliance with standards (CWS) element describes a process for maintaining adherence to applicable codes, standards, regulations, and laws (*standards*), the attributes of a standards system, and the steps an organization might take to implement the CWS element.

Note: In this book, we are using the term "standards" generically and broadly to apply to applicable.

- **Codes.** Codes typically are definitively focused, generally prescriptive, have safety embedded, and usually focus on new construction. In the United States, they are issued by bodies such as the American Society of Mechanical Engineers (ASME), ANSI, the American Society for Testing and Materials (ASTM) International, the National Board of Boiler and Pressure Vessel Inspectors, and the National Fire Protection Association (NFPA).
- **Standards.** Typically, standards also are definitively focused and have safety embedded, but require more interpretation than codes. They are

generally issued by industry organizations such as API, the Chlorine Institute (CI), the International Institute of Ammonia Refrigeration (IIAR), and the Technical Association of the Pulp and Paper Industry (TAPPI).

- **Regulations/laws.** These typically result either from government agencies including code-type requirements in regulations/laws or citing specific existing codes (thereby making them part of the law).

CWS requires a system to identify, develop, acquire, evaluate, disseminate, and provide access to applicable codes, standards, regulations, and laws that affect process safety. The standards system addresses both internal and external standards; national and international codes and standards; and local, state, and federal regulations and laws. The system makes this information easily and quickly accessible to potential users. The standards system interacts in some fashion with every RBPS management system element.

If the CWS element work is done at a company level, then the responsible party keeps a list of all applicable requirements and copies of all such updated documents. This information is typically communicated to division- and facility-level personnel responsible for local compliance activities.

Facilities that exhibit a high demand rate for maintaining compliance with frequently changing standards may need greater specificity in the standards procedure and larger allocation of personnel resources to fulfill the defined roles and responsibilities. Lower demand situations can allow facilities to operate a standards protocol with greater flexibility – possibly with a single person providing the advisory service at a divisional or corporate level for multiple facilities. Facilities with strong process safety cultures generally will have more performance-based standards procedures, allowing trained employees to use good judgment in managing compliance. Facilities with an immature or evolving process safety culture may require more prescriptive standards procedures, more frequent auditing, and greater command and control management system features to ensure good standards implementation discipline.

Another common term important to the CWS element is recognized and generally accepted good engineering practices (RAGAGEPs). A RAGAGEP is a consensus code, standard, or guideline that provides the engineering practices for the design, fabrication, installation, maintenance, and/or inspection/testing of equipment. A RAGAGEP can be mandatory (e.g., regulatory or insurance), suggested, or common practice. Without a good RAGAGEP program, the effectiveness of compliance efforts and the safety of operations may be reduced.

Following are steps to ensure that thorough consideration of the applicable and appropriate RAGAGEPs has taken place:

1. Develop a general knowledge of RAGAGEPs.

2. Identify which RAGAGEPs may apply to your equipment, chemicals, or processes.
3. Assess the applicability and scope of each candidate RAGAGEP.
4. Document the approach.

The need for documentation cannot be emphasized too strongly. Proper documentation demonstrates that you have undertaken and completed a thorough consideration of the appropriate RAGAGEPs. The objective is to develop enough understanding to answer the following questions:

- Which organizations issue RAGAGEPs applicable to the industry and type of facility?
- Which organizations issue RAGAGEPs applicable to the chemicals or hydrocarbons handled at your facility?
- Which organizations issue RAGAGEPs applicable to the types of equipment in your facility?
- What are the scopes and applications of these organizations' RAGAGEPs?

Keep in mind that standards often provide alternate methods and frequently change, so applying some of the RAGAGEPs will require interpretation and judgment.

Common challenges for a RAGAGEP program, particularly in small facilities, include:

- developing RAGAGEP awareness among staff members,
- developing and maintaining a RAGAGEP knowledge base,
- dedicating the resources required to implement applicable RAGAGEPs,
- dealing with incomplete or no documentation (e.g., of equipment design or fabrication), and
- dealing with equipment not in compliance with RAGAGEPs.

When resources are severely limited, maintaining and using the level of expertise required to meet common accepted standards is a major challenge.

Once a CWS element and/or RAGAGEP program knowledge base has been established, it must be maintained. One common problem is losing this knowledge base when personnel change assignments or leave the company. This is particularly a problem for smaller facilities where only one or two people serve as the RAGAGEP experts. The following methods can be utilized to assist in maintaining the program:

- Incorporate ongoing awareness of applicable RAGAGEPs in written job descriptions, goals, objectives, and performance appraisals.
- Include RAGAGEPs in the initial training agenda for new employees.

- Instead of developing a single expert, develop experts in each engineering discipline.
- Provide opportunities for different individuals to attend RAGAGEP training courses.
- Develop a network of RAGAGEP experts from the corporate engineering department and local contractors.

Finally, consider striving for proactive compliance in your CWS element or RAGAGEP program. Proactive compliance can be defined as "staying ready so you do not have to get ready." It involves using the wide body of knowledge that is embedded in the codes, standards, and RAGAGEPs to (1) assist with the conduct of business in the most cost-effective and efficient manner and (2) help avoid losses. Proactive compliance sets the attitude that you will use a code to achieve a high level of accomplishment and achievement. It forces you to determine a positive course of action and then implement it.

The "Mechanical Integrity Test and Inspection (MITI) Guide" on the files on the Web accompanying this book (provided by Eli Lilly and Company) provides an example of implementing the CWS element using RAGAGEPs.

5.3.3 Process Safety Competency

Developing and maintaining a Process safety competency (PSC) element encompasses three interrelated actions: (1) continuously improving knowledge and competency, (2) ensuring that appropriate information is available to people who need it, and (3) consistently applying what has been learned.

The learning aspect includes efforts to develop, discover, or otherwise enhance knowledge. It ranges from narrowly defined tasks that develop new information based on a specific request, such as conducting experiments that provide data needed by hazard identification and risk analysis teams, to wide-ranging efforts to maintain and advance the knowledge base of the entire organization or even a sector of the chemical industry. The learning aspect also includes structured means to retain people-based knowledge, including succession planning.

The PSC element is closely related to the knowledge and training elements of the RBPS system. While the PSC element often generates new information, the knowledge element provides the means to catalog and store information so that it can be retrieved on request. The PSC element focuses primarily on organizational learning, whereas the training element addresses efforts to develop and maintain the competence of each individual worker.

The PSC element involves increasing the body of knowledge and, when applicable, pushing newly acquired knowledge out to appropriate parts of the organization, sometimes independently of any request. Most importantly, this element supports the application of this body of process knowledge to situations that help manage risk and improve facility performance.

Unlike some of the other RBPS elements, no simple answer exists as to "how" PSC is achieved. The single most important factor is a commitment by senior management to support efforts to learn and to share new information and insights among units at a facility, with sister facilities within the company, and potentially with other companies. Once the commitment is in place, opportunities to learn and interact with others abound. Some information will have to be passed along through mentorship and collaboration; both of these activities typically require active management support to ensure success. A closely related activity is succession planning, which is an intentional activity that helps ensure that key positions are staffed with individuals who possess specific knowledge and experience.

Note: "Vision 20/20," developed by CCPS (Ref. 5.5), looks into the not-too-distant future to describe how great process safety is delivered when it is collectively and fervently supported by industry, regulators, academia, and the community worldwide; driven by the five industry tenets; and enhanced by the four global societal themes. One of the five industry tenets is "Intentional Competency Development to ensure that all employees who impact process safety are fully capable of meeting the technical and behavioral requirements for their jobs." The bottom line: No matter how good the culture or management system is, or how well the company adheres to standards, it takes highly competent employees to implement those systems or standards. And that requires intentional competency development.

*Note: CCPS Project 239 (*Guidelines for Process Safety Knowledge and Expertise*) will specifically address establishing and maintaining PSC within organizations.*

5.3.4 Stakeholder Outreach

Stakeholder outreach (outreach) is a process for:

- seeking out individuals or organizations that can be, or believe they can be, affected by company operations and engaging them in a dialogue about process safety;
- establishing a relationship with community organizations, other companies, professional groups, and local, state, and federal authorities; and
- providing accurate information about the company's and the facility's products, processes, plans, hazards, and risks.

This process ensures that management makes relevant PSI available to a variety of organizations. This element also encourages the sharing of relevant information and lessons learned with similar facilities within the company and with other companies in the industry group. Finally, the outreach element promotes involvement of the facility in the local community and facilitates communication of information and facility activities that could affect the community.

Companies train key personnel to interact with important stakeholder groups during planned events and provide resources for all employees to use in their everyday encounters with the public. Crisis communication and outreach training is provided to senior management to help deal with episodic events.

Higher risk situations usually dictate a greater need for formality and thoroughness in the implementation of the outreach element. Conversely, companies having lower risk situations may appropriately decide to pursue outreach activities in a less rigorous fashion. In the case of the outreach element, risk takes on a two-fold meaning: (1) the risk of experiencing an incident and (2) the risk of experiencing an adverse stakeholder reaction as a result of a process safety issue at the facility or other facilities within the company or industry. A higher risk situation may demand a more formal risk communication program that provides detailed information to stakeholders and keeps them updated. In a lower risk situation, a general community outreach policy, via informal practices by trained key employees, may be sufficient.

5.3.5 Conduct of Operations (Operational Discipline)

Conduct of operations (COO) is the execution of operational and management tasks in a deliberate and structured manner. It is also sometimes called operational discipline (OD) or formality of operations, and it is closely tied to an organization's culture. Conduct of operations institutionalizes the pursuit of excellence in the performance of every task and minimizes variations in performance. Workers at every level are expected to perform their duties with alertness, due thought, full knowledge, sound judgment, and a proper sense of pride and accountability (Refs. 5.6 and 5.7).

To develop an effective COO element program, an organization must start with an honest statement of its objectives and risk tolerance. Considering the outputs of other elements, the organization can then formulate an operations policy and document it, along with the procedures for its implementation. However, the program cannot be merely words on paper. Workers must be trained on the policies and procedures so that they understand the goals and expectations, the lines of authority, and their personal accountability. They must apply good reasoning and judgment (founded upon a sound process safety culture) in all situations, but particularly when action is required in situations not specifically addressed by policy or procedure.

Beyond that, the most critical, ongoing requirement is that management lead by example. If a procedure instructs workers to shut down the process under defined emergency conditions, but management praises operators who "ride it out" and avoid a shutdown, then operational discipline will suffer. COO and OD tolerate no deviation from approved procedures, even if the outcome of a deviation is inconsequential or desirable. Thus, management must hold workers accountable for their actions in all circumstances to avoid the normalization of deviation.

The CCPS book *Conduct of Operations and Operational Discipline for Improving Process Safety in Industry* provides extensive discussion and guidance on this element. In particular, Chapter 7 provides guidance on implementing and maintaining COO/OD systems. The key points from Chapter 7 are summarized in the Plan-Do-Check-Adjust (PDCA) process shown in Table 5.1.

Table 5.1 The PDCA Process Applied to COO/OD Implementation

PLAN: Analyze the situation and develop a plan	• Set a measurable objective toward the goal for the COO/OD effort • Identify the processes impacted by COO/OD • Select where to apply COO/OD • List the steps in each process as it currently exists • Map each process • Identify issues related to COO/OD implementation • Collect data on the current process • Generate implementation plans • Gain approval and support
DO: Implement the plan	• Implement the chosen solution on a trial or pilot basis (first pass through the PDCA cycle) • Implement the change throughout the organization (subsequent passes through the PDCA cycle)
CHECK: Evaluate the results	• Gather data on the modified system results • Analyze the results data • Achieved the desired goal? o If YES, skip the Act step, revise the goal to the next objective for continuous improvement, update the plan, and repeat the PDCA cycle o If NO, proceed to the Act step, modify the implementation plan, and repeat the cycle
ADJUST: Standardize the implementation (and continuously improve)	• Identify systemic changes and training needs for full implementation • Plan ongoing monitoring of the COO/OD system • Continue to look for incremental improvements to COO/OD

5.3.6 Measurement and Metrics

The measurement and metrics (metrics) element establishes performance and efficiency indicators to monitor the near-real-time effectiveness of the RBPS management system and its constituent elements and work activities (Ref. 5.8). This element addresses which indicators to consider, how often to collect data, and what to do with the information to help ensure responsive, effective RBPS management system operation.

A combination of leading and lagging indicators is often the best way to provide a complete picture of process safety effectiveness (Ref. 5.9). Outcome-oriented lagging indicators, such as incident rates, are generally not sensitive enough to be useful for continuous improvement of PSM systems because incidents occur too infrequently. Measuring PSM performance requires the use of leading indicators, such as the rate of improperly performed line-breaking activities.

Metrics can be established as a facility designs, corrects, or improves its PSM system (Ref. 5.10). Establishing metrics (in particular, the data-gathering and refreshing mechanisms) is simpler to do during the initial design and implementation of the system. Each RBPS book chapter has a section that contains a list of possible metrics proposed for that element's key principles (Section X.5, where X is the chapter number). Readers can select from these examples or develop their own ideas. Typically, a small set of metrics is proposed, data are gathered, and the set is pilot tested to see if tracking the metric data helps identify management system degradation. This metrics experiment should last a minimum of several "metric refresh cycles" and, at most, until the next formal RBPS audit is conducted. At that time, the audit can show whether the metrics have been correctly projecting the performance of the PSM system.

Extensive additional information and guidance regarding measurement and metrics can be found in the following resources:

- A CCPS publication entitled *Process Safety Leading and Lagging Metrics . . . You Don't Improve What You Don't Measure* (Ref. 5.11). It describes three categories of metrics:
    - **Lagging metrics.** Lagging metrics are a retrospective set of metrics that are based on incidents that meet the threshold of severity that should be reported as part of the industry-wide process safety metric.
    - **Leading metrics.** Leading metrics are a forward-looking set of metrics that indicate the performance of the key work processes, operating discipline, or layers of protection that prevent incidents
    - **Near-miss and other internal lagging metrics.** These metrics describe less severe incidents (i.e., below the threshold for inclusion in the industry lagging metric), or unsafe conditions that activated one or more layers of protection. Although these are actual events (i.e., lagging metrics), they are generally considered to be good indications of conditions that could ultimately lead to a more severe incident.

The document strongly recommends that all companies incorporate each of these three types of metrics into their internal PSM system. Recommended metrics for each of these categories are included in the three primary sections of the document.

- The CCPS book *Guidelines for Process Safety Metrics* (Ref. 5.12). This book provides basic information on process safety performance indicators, including a comprehensive list of metrics for measuring performance and examples of how they can be successfully applied over both the short and long term. Readers can use the guidance in this book to help identify appropriate metrics useful in monitoring performance and improving process safety programs.
- The CCPS book *Integrating Management Systems and Metrics to Improve Process Safety Performance* (Ref. 5.13). This book was written, in part, to capture the recent advances in understanding how process safety performance improvements can be measured with a combination of leading and lagging indicators. Since the management programs for the process and personal safety, health, environment, quality, and security (SHEQ&S) groups have developed separately in many organizations, these guidelines were written to help organizations identify common process safety metrics across the SHEQ&S groups. Integrating these metrics will reduce an organization's overall operational risks.
- *Developing Process Safety Indicators: A Step-by-Step Guide for Chemical and Major Hazard Industries* (Ref. 5.10). This Health and Safety Executive publication presents a six-step process for developing and implementing safety indicators: (1) establish the organizational arrangements to implement indicators, (2) decide on the scope of the indicators, (3) identify the risk control systems and decide on the outcomes, (4) identify critical elements of each risk control system, (5) establish a data-collection and reporting system, and (6) review.
- The Organisation for Economic Cooperation and Development (OECD) publication *Guidance for Industry, Public Authorities and Communities for Developing SPI Programmes Related to Chemical Accident Prevention, Preparedness and Response* (Ref. 5.9). The three chapters in this document are designed to help public authorities (including emergency response personnel) and organizations representing communities/public better understand safety performance indicators (SPI) and how to implement SPI programs. However, many of the approaches suggested can also be applied to implementing metrics programs in the petrochemical industry.
- ANSI/API RP 754, *Process Safety Performance Indicators for the Refining and Petrochemical Industries* (Ref. 5.14). This recommended practice describes possible metrics that are grouped into four tiers and address both leading and lagging indicators.

Note: Integrating Management Systems and Metrics to Improve Process Safety Performance *provides examples of implementing this element.*

5.3.7 Management Review and Continuous Improvement

Management review is the routine evaluation of whether management systems are performing as intended and producing the desired results as efficiently as possible. It is the ongoing due diligence review by management that fills the gap between day-to-day work activities and periodic formal audits, thereby allowing ongoing evaluation and guiding continuous improvement. Management reviews have many of the characteristics of a first-party audit as described in Chapter 21 of the RBPS book. They require a similar system for scheduling, staffing, and effectively evaluating all RBPS elements, and a system should be in place for implementing any resulting plans for improvement or corrective action and verifying their effectiveness.

Management reviews are conducted with the same underlying intent as an audit – to evaluate the effectiveness of the implementation of an entire RBPS element or a particular element task. However, because the objective of a management review is to spot current or incipient deficiencies, the reviews are more broadly focused and more frequent than audits, and they are typically conducted in a less formal manner.

Nevertheless, like an audit a management review should at least check the implementation status of one or more RBPS elements against established requirements. The management review team meets with the individuals responsible for managing and executing the subject element to (1) present program documentation and implementation records, (2) offer direct observations of conditions and activities, and (3) answer questions about program activities. The team attempts to answer such questions as:

- What is the quality of our program?
- Are these the results we want?
- Are we working on the right things?

Organizational changes, staff changes, new projects or standards, efficiency improvements, and any other anticipated challenges to the subject element are also discussed so that management can proactively address those issues.

Recommendations for addressing any existing or anticipated performance gaps or inefficiencies are proposed, and responsibilities and schedules for addressing the recommendations are assigned. Typically, the same system used to track corrective actions from audit findings is used to track management review recommendations to their resolution. The meeting minutes and documentation of each recommendation's resolution are maintained as required to meet programmatic needs.

Management review results should be monitored over time, and more frequent reviews should be scheduled if persistent problems are evident.

5.4 MONITORING NEW ELEMENTS OR ACTIVITIES

Section 4.6 of this book discusses monitoring the initial implementation and performance of a new PSM system. Similarly, in addition to establishing ongoing measurement, management review, and auditing requirements for the new elements (as discussed in Section 4.4), also consider the need to:

- establish a plan/means for collecting PSM metrics and other feedback and data on the performance of the new elements,
- develop a plan for the initial conduct of PSM management reviews, and
- create a plan for the initial auditing of the PSM elements/activities.

Note: The guidance provided in Section 4.6 regarding each of these activities can be utilized.

5.5 REFERENCES

5.1 Jones, David, "Turning the Titanic – Three Case Histories in Cultural Change," CCPS International Conference and Workshop, Toronto, Ontario, Canada, 2001.

5.2 Schein, Edgar H., *Organizational Culture and Leadership, 3rd Edition,* Jossey-Bass, San Francisco, California, 2004.

5.3 Energy Institute, "Winning Hearts and Minds," London, England, UK, www.eimicrosites.org/heartsandminds/index.php.

5.4 Lodal, Peter, "Take the FUN Out of Process Safety," *CEP Magazine*, American Institute of Chemical Engineers, February 2014, www.aiche.org/sites/default/files/cep/20140226_1.pdf.

5.5 Center for Chemical Process Safety of the American Institute of Chemical Engineers, "Vision 20/20," New York, New York, www.aiche.org/ccps/resources/vision-2020.

5.6 International Atomic Energy Agency, *Conduct of Operations at Nuclear Power Plants Safety Guide*, Vienna, Austria, 2008.

5.7 U.S. Department of Energy, *Conduct of Operations Requirements for DOE Facilities*, DOE Order 5480.19 Chg 2, Washington D.C., 2001, www.directives.doe.gov/directives-documents/5400-series/5480.19-BOrder-chg2.

5.8 Center for Chemical Process Safety of the American Institute of Chemical Engineers, "ProSmart – The Tool You Need to Improve Process Safety," New York, New York, www.aiche.org/ccps/prosmart/.

5.9 Organisation for Economic Co-operation and Development, *Guidance on Safety Performance Indicators: Guidance for Industry, Public Authorities and Communities for Developing SPI Programmes Related to Chemical Accident Prevention, Preparedness and Response*, OECD Environment, Health and Safety Publications, Series on Chemical Accidents, No. 11, Danvers, Massachusetts, 2005, www.oecd.org/chemicalsafety/chemical-accidents/.

5.10 Health and Safety Executive, *Developing Process Safety Indicators: A Step-by-Step Guide for Chemical and Major Hazard Industries*, Norwich, England, UK, 2006, www.hse.gov.uk/pubns/books/hsg254.htm.

5.11 Center for Chemical Process Safety of the American Institute of Chemical Engineers, *Process Safety Leading and Lagging Metrics . . . You Don't Improve What You Don't Measure,* New York, New York, January 2011, www.aiche.org/sites/default/files/docs/pages/CCPS_ProcessSafety_Lagging_2011_2-24.pdf.

5.12 Center for Chemical Process Safety of the American Institute of Chemical Engineers, *Guidelines for Process Safety Metrics*, John Wiley & Sons, Inc., Hoboken, New Jersey, 2009.

5.13 Center for Chemical Process Safety of the American Institute of Chemical Engineers, *Guidelines for Integrating Management Systems and Metrics to Improve Process Safety Performance*, John Wiley & Sons, Inc., Hoboken, New Jersey, 2015.

5.14 ANSI/API Recommended Practice 754, *Process Safety Performance Indicators for the Refining and Petrochemical Industries*, American Petroleum Institute, Washington D.C., April 2010, www.publications.api.org/.

6

IMPROVING AN EXISTING PSM ELEMENT OR SYSTEM

6.1 DETERMINING WHICH ELEMENTS TO IMPROVE

Improving a PSM element or system (program) will be difficult without the appropriate management support to do so. The recommended first step in gaining management support is to conduct a high-level value gap analysis of the entire PSM program. This analysis will not only identify compliance gaps in the company's PSM program, it will also show gaps in efficiency, general safety, uptime, and cost. Of course, companies may choose to do only what is necessary to meet compliance versus seeking "extra" value through cost avoidance, improved facility efficiencies, etc.

Note: This chapter focuses on the improvement of an existing PSM element or system at one site. Multinational organizations attempting to implement similar actions at multiple sites will face additional challenges.

6.1.1 Value Gap Analysis

The initial step in improving an existing PSM element or system is to determine which elements or parts of the system require improvement. There are certainly a number of ways this can be done, but the most effective is normally to conduct a value gap analysis. A determination of the value for variances identified in the program will be the basis for justification to proceed with improvements to the PSM program. The value gap analysis can be developed from the last PSM audit results or a separate PSM assessment. It will include the value of improved safety performance, increased facility utilization or uptime, cost avoidance, and regulatory compliance. The PSM coordinator typically facilitates the value gap analysis with significant input from engineering, maintenance, and operations personnel. Once developed, the value gap analysis is then reviewed with the various levels of management required to gain approval to proceed. An example of a high-level value gap analysis for some of the PSM elements is shown in Table 6.1.

Table 6.1 Example of a High-level Value Gap Analysis

PSM Requirement – PHA	PSM Element Gap	Value of Gap	Recommendation to Close Gap
[1910.119(e)(3)]. The process hazard analysis shall address: (i) The hazards of the process; (ii) The identification of any previous incident which had a likely potential for catastrophic consequences in the workplace; (iii) Engineering and administrative controls applicable to the hazards and their interrelationships such as appropriate application of detection methodologies to provide early warning of releases. (Acceptable detection methods might include process monitoring and control instrumentation with alarms, and detection hardware such as hydrocarbon sensors); (iv) Consequences of failure of engineering and administrative controls; (v) Facility siting; (vi) Human factors; and (vii) A qualitative evaluation of a range of the possible safety and health effects of failure of controls on employees in the workplace.	Current PHAs do not address previous incidents that had a likely potential for catastrophic consequences.	Closing this gap will improve safety of the process by identifying hazardous scenarios that have occurred, allowing the PHA team to consider these incidents when assessing the adequacy of existing adverse reaction QRAs. Closing this gap will also ensure regulatory compliance. Closing this gap will also prevent costs associated with possible incidents that may have arisen as a result of not addressing potential hazards (determine the cost of a serious incident in the process that included loss of containment and potential injuries).	Include a review of previous significant incidents and near misses in all future PHAs.

Table 6.1 ***Continued***

PSM Requirement – Operating Procedures	PSM Element Gap	Value of Gap	Recommendation to Close Gap
(ii) Operating limits: (A) Consequences of deviation; and (B) Steps required to correct or avoid deviation.	Consequences of deviation and steps required to correct the deviation are not readily available to operations personnel.	Closing this gap will improve safety of the process by providing a vital tool to operations personnel to use during training for response to upset or emergency conditions. Closing this gap will also ensure regulatory compliance. Closing this gap will also prevent costs associated with process upsets that may have been prevented if this requirement had been in place. (determine the cost of a process upset).	Update operating procedures to include consequences of deviation to critical process variables and include steps required to correct or avoid the deviation.

Table 6.1 *Continued*

PSM Requirement – Mechanical Integrity	PSM Element Gap	Value of Gap	Recommendation to Close Gap
[1910.119(j)(4)]. Inspection and testing. (i) Inspections and tests shall be performed on process equipment. (ii) Inspection and testing procedures shall follow recognized and generally accepted good engineering practices. (iii) The frequency of inspections and tests of process equipment shall be consistent with applicable manufacturer's recommendations and good engineering practices, and more frequently if determined to be necessary by prior operating experience. (iv) The employer shall document each inspection and test that has been performed on process equipment. The documentation shall identify the date of the inspection or test, the name of the person who performed the inspection or test, the serial number or other identifier of the equipment on which the inspection or test was performed, a description of the inspection or test performed, and the results of the inspection or test.	Inspections and tests have not been routinely conducted on process piping from the feed tank to the reactors, nor have they been conducted on the reactors.	Closing this gap will improve safety of the process as the data will allow identification of required repairs or replacement before a loss of containment occurs. Closing this gap will also ensure regulatory compliance. Closing this gap will prevent costs associated with equipment downtime and a loss of containment incident (determine the cost of equipment downtime and a loss of containment incident).	Establish an inspection, test, and preventive maintenance (ITPM) program for the reactor feed piping and associated reactors.

Table 6.1 *Continued*

PSM Requirement – Hot Work Permit	PSM Element Gap	Value of Gap	Recommendation to Close Gap
[1910.119(k)(2)]. The permit shall document that the fire prevention and protection requirements in 29 CFR 1910.252(a) have been implemented prior to beginning the hot work operations; it shall indicate the date(s) authorized for hot work; and identify the object on which hot work is to be performed.	The identity of the object on which hot work will be performed is not routinely included on hot work permits.	Closing this gap will improve safety of the process as each permit will clearly state where hot work will be performed. Closing this gap will also ensure regulatory compliance. Closing this gap will prevent costs associated with a potential fire/hazardous chemical release or injuries (determine the cost of a process fire/hazardous chemical release with injuries).	Establish a process to verify that the object on which welding is to occur is clearly identified on all hot work permits. Train personnel on properly completing hot work permits. Conduct periodic audits to ensure compliance with this requirement.

Eli Lilly and Company (Lilly), a global pharmaceutical firm, went through an effort to improve its existing PSM system. This improvement effort included developing and completing a "Brown Paper" Should-Be/As-Is Gap Analysis process. The Brown Paper process involved placing large rolls of brown paper/butcher paper on walls and writing the Should-Be map on the paper. The gap analysis team then worked with site teams to map the business process steps at each plant site. When gaps were identified, an action plan was developed to address each gap. Lilly also sent implementation teams of at least three people to each plant site to conduct As-Is/Should-Be Gap Analyses at these sites. Each site was required to supply at least three teams for up to four weeks at a time, with the purpose of creating action plans with sites that would integrate Lilly's internal PSM process – Globally Integrated Process Safety Management (GIPSM) – with local site processes, owned by local personnel. See Appendix II for more information on Lilly's PSM implementation effort and this gap analysis.

The value gap analysis will also be used later when identifying the areas of PSM that should be assessed as described in Section 6.2.

As is typical for project development, a cost and resource analysis should be developed to accompany the justification. Most of the costs will be to provide the employees required to staff the project. A full-time leader will be required, as will a team of three to seven employees, depending on the scope of the project. Time allocation requirements for individual team members may range from 10 percent to 25 percent.

6.2 ASSESSING THE PROGRAM AND DETERMINING THE ROOT CAUSES OF POOR PERFORMANCE

6.2.1 Program Assessment

Once management agrees with the justification, cost, and resource loading for the project, an overall assessment of the PSM program should be conducted to initiate improvement for a PSM element or system. To obtain an unbiased assessment of the various elements of the program, an independent third party will often be involved in this review. For example, the assessment could be completely conducted by a third party, led by a third party, or include a third-party team member.

The assessment should be comprehensive and not just a sampling of the selected PSM elements or system. In other words, it should be more extensive than an analysis of variance from the PSM standard. It should also be a benchmarking exercise versus similar facilities in the same company or similar facilities in the same industry. A third party may provide significant benefits here as they bring knowledge of what is done at other sites and other companies. Benchmarking data can be very difficult to obtain on an intercompany basis.

The assessment will be similar in nature to a PSM audit, but it will delve deeper and will point out where best practices could be implemented to enhance programs that already meet regulatory compliance. One key to the assessment will be sharing of ideas by experienced facility personnel. If a certain area of PSM is not working as intended, facility personnel can help shed light on the issues.

A standard PSM audit table may be used to document the results of the assessment. Tasks that should be completed during the assessment include:

- employees' roundtable discussion of PSM issues,
- best practice sharing by the site and the assessment team,
- records reviews,
- review of previous audits,
- employee interviews,
- review of corrective action follow-up systems currently in use,
- field tours, and
- review of PSM elements/systems in action (MOC, PSSR, safe work permits, PHA, inspections, etc.).

See Table 6.2 for an example PSM assessment format. This is an abbreviation of a table listing the 14 OSHA PSM elements in the first column. A completed assessment table would typically list all of the requirements of each element per the PSM standard or other framework being assessed in the first column. The assessment team would utilize/consider each task for all the requirements of each element. The company would decide whether recommendations are to be included in the assessment. The PSM program improvement team may be asked to develop the recommendations instead of the assessment team.

A company may decide to limit the scope of this PSM assessment based on the outcome of the value gap analysis described in section 6.1.1. The value gap analysis is a tool that can be used to set priorities for the PSM program improvement team. The more significant the gap (safety, regulatory, cost), the higher the priority will be. Using this tool to set priorities is especially important when resources are limited and competing with other important initiatives.

6.2.2 Root Causes of Performance Gaps

A company may choose different ways to determine the root cause(s) of the gap(s) that are identified. A performance gap from the assessment is normally an indication of a management system weakness. This means that the root cause was something over which management had control. Figure 6.1 presents a typical flowchart for determining root causes.

In this flowchart a loss event or condition represents the performance gap. The causal factors are the equipment or personnel performance gaps.

Table 6.2 Example PSM Assessment Protocol

Element	PSM Assessment Tasks	Assessment Results	Gap Versus Standard/ Best Practice
[1910.119(c)(1)]. *Employee Participation (EP)*	1. Employees roundtable discussion of PSM issues – Are employees satisfied with the plan? 2. Best practice sharing by the site and the assessment team – Does the plan cover more than just the 14 regulatory elements? What sharing does the assessment team have for the site? 3. Records reviews – Is the plan up to date with current practice? 4. Review of previous audits – Have there been any issues cited with the plan in the past? 5. Employee interviews – Any other issues not surfaced in the roundtable? 6. Review of follow-up systems in use – N/A 7. Field tours – N/A 8. Review of PSM elements in action (MOC, PSSR, safe work permits, PHA, inspections, etc.) – Are employees actively involved in all elements?		

Table 6.2 *Continued*

Element	PSM Assessment Tasks	Assessment Results	Gap Versus Standard/ Best Practice
[1910.119(d)]. *Process Safety Information (PSI)*	1. Employees roundtable discussion of PSM issues – Are employees satisfied with the completeness of the PSI? What gaps are there in the PSI? Is the PSI readily available? 2. Best practice sharing by the site and the assessment team – Does the PSI go further than the standard? What are the systems by which employees can access the PSI? Are these systems user friendly? What sharing does the assessment team have for the site? 3. Records reviews – Has the PSI been kept current? Is the PSI complete? 4. Review of previous audits – Have there been any issues cited with the PSI in the past? 5. Employee interviews – Any other issues not surfaced in the roundtable? 6. Review of follow-up systems in use – What systems are used to keep track of drawing updates, SDS updates, file updates, etc.? 7. Field tours – Are P&IDs up to date with what is in the field? 8. Review of PSM elements in action (MOC, PSSR, safe work permits, PHA, inspections, etc.) – Is PSI updated in conjunction with MOCs? Is PSI available and up to date for use in PHAs, inspections, etc.?		

Table 6.2 *Continued*

Element	PSM Assessment Tasks	Assessment Results	Gap Versus Standard/ Best Practice
[1910.119(e)(1)]. Process Hazard Analysis (PHA)	1. Employees roundtable discussion of PSM issues – Are employees satisfied with how PHAs are conducted? Are employees adequately involved in all PHAs? Are the results of all PHAs shared with all employees? Do employees believe recommendations from PHAs are implemented within an appropriate time frame? 2. Best practice sharing by the site and the assessment team – Do PHA scopes go above and beyond the standard? Are more employees used on PHA teams than what the standard calls for? Are procedural PHAs carried out? What sharing does the assessment team have for the site? 3. Records reviews – Have PHAs been conducted within the five-year windows? Is PHA content in compliance with the standard? Have recommendations been resolved and implemented in a timely manner? 4. Review of previous audits – Have there been any issues cited with the PHA in the past? 5. Employee interviews – Any other issues not surfaced in the roundtable? 6. Review of follow-up systems in use – What systems are used to keep track of PHA recommendations? 7. Field tours – Are PHA scenarios used in the control rooms on a daily basis? 8. Review of PSM elements in action (MOC, PSSR, safe work permits, PHA, inspections, etc.) – Sit in on a PHA? Ask operations personnel about PHA hazard scenarios.		

Table 6.2 *Continued*

Element	PSM Assessment Tasks	Assessment Results	Gap Versus Standard/ Best Practice
[1910.119(f)(1)]. Operating Procedures (OP)	1. Employees roundtable discussion of PSM issues – Are employees satisfied with the completeness of process operating procedures? Are operating procedures clear and easy to understand? Do employees have an opportunity to make recommendations to update operating procedures when more efficient ways to operate are discovered? 2. Best practice sharing by the site and the assessment team – Do operating procedures cover more than what the standard requires? Are operating procedures laid out in checklist format? What systems are used for employees to access operating procedures? Are these systems user friendly? Are there safe work practices for lockout/tagout, confined space entry, opening process equipment/piping, and control of entrance to and exit from processes? What sharing does the assessment team have for the site? 3. Records reviews – Have the procedures been kept current? Are the procedures complete? Review a cross section of completed work permits for safe work practices. 4. Review of previous audits – Have there been any issues cited with the operating procedures in the past? 5. Employee interviews – Any other issues not surfaced in the roundtable? 6. Review of follow-up systems in use – What systems are used to keep track of procedure updates? New procedures? 7. Field tours – Are operating procedures being used out in the process? 8. Review of PSM elements in action (MOC, PSSR, safe work permits, PHA, inspections, etc.) – Access to procedures in the control rooms.		

Table 6.2 *Continued*

Element	PSM Assessment Tasks	Assessment Results	Gap Versus Standard/ Best Practice
[1910.119(g)(1)]. *Training (TRN)*	1. Employees roundtable discussion of PSM issues – Are employees satisfied with the training program? Are testing protocols adequate to ensure understanding of the material presented? Do employees have an opportunity to make recommendations to improve the training program? 2. Best practice sharing by the site and the assessment team – Do current training programs cover more than what the standard requires? What types of electronic systems are used for training programs? Are these systems user friendly? What sharing does the assessment team have for the site? 3. Records reviews – Is all refresher training current? Is new hire training complete according to the site's training protocol? Does the site train on operating procedures every three years as required? 4. Review of previous audits – Have there been any issues cited with training in the past? 5. Employee interviews – Any other issues not surfaced in the roundtable? 6. Review of follow-up systems in use – What systems are used to keep track of upcoming training? Of overdue training courses? 7. Field tours – Is there any training underway during the assessment? Classroom? On-the-job training? 8. Review of PSM elements in action (MOC, PSSR, safe work permits, PHA, inspections, etc.) – Training program changes in the works?		

Table 6.2 *Continued*

Element	PSM Assessment Tasks	Assessment Results	Gap Versus Standard/ Best Practice
[1910.119(h)(1)]. *Contractors (CON)*	1. Employees roundtable discussion of PSM issues – Are contractors well trained to work in PSM areas? Do contractors understand PSM? Is safety awareness for contractors equal to that of facility employees? 2. Best practice sharing by the site and the assessment team – Do contractors exercise stop work authority? Do contractors introduce best practice methods from other sites where they work? What sharing does the assessment team have for the site? 3. Records reviews – Are training records up to date for each contract employee? Do contractors follow set training programs with established testing protocols? Do contractor safety metrics meet company requirements? 4. Review of previous audits – Have there been any issues cited with contractors in the past? 5. Employee interviews – Any other issues not surfaced in the roundtable? 6. Review of follow-up systems in use – What systems are used to keep track of contractor training? Of contractor safety metrics? What systems are used to track contractor performance? 7. Field tours – Are audits performed of contractor safe work practices? Are audits performed off site at contractor shop locations? 8. Review of PSM elements in action (MOC, PSSR, safe work permits, PHA, inspections, etc.) – Are contractors following all permit requirements, including signing in to PSM areas?		

Table 6.2 *Continued*

Element	PSM Assessment Tasks	Assessment Results	Gap Versus Standard/ Best Practice
[1910.119(i)(1)]. Pre-Startup Safety Review (PSSR)	1. Employees roundtable discussion of PSM issues – Are employees involved in PSSRs? Are actions from PSSRs completed before startup (when required)? Are the voices of employees heard during PSSRs? Is all training completed before startup (when required)? 2. Best practice sharing by the site and the assessment team – Is there a routine audit in place to ensure compliance with the site PSSR procedure? What type of database is used to track PSSRs? Is this system user friendly? What sharing does the assessment team have for the site? 3. Records reviews – Has the site been following its PSSR procedure? Are PSSR actions and training completed in a timely manner? Is there a PSSR for every MOC? 4. Review of previous audits – Have there been any issues cited with PSSRs in the past? 5. Employee interviews – Any other issues not surfaced in the roundtable? 6. Review of follow-up systems in use – What systems are used to keep track of PSSRs and PSSR actions? What happens when actions are not completed in a timely manner? 7. Field tours – Pull a completed PSSR and check the field installation versus the MOC. 8. Review of PSM elements in action (MOC, PSSR, safe work permits, PHA, inspections, etc.) – Witness a PSSR in the field.		

Table 6.2 *Continued*

Element	PSM Assessment Tasks	Assessment Results	Gap Versus Standard/ Best Practice
[1910.119(j)(1)]. Mechanical Integrity (MI)	1. Employees roundtable discussion of PSM issues – Is there an active ITPM program for the site? Does the ITPM program encompass everything that is required per the industry or API standard that the site follows? What protocols are used to determine when to repair versus replace equipment? 2. Best practice sharing by the site and the assessment team – Does the site go above and beyond what the industry or API standard requires? Is the site using risk based inspection (RBI)? Does the site make conservative decisions for MI questions? What sharing does the assessment team have for the site? 3. Records reviews – Are there overdue inspections of process piping and vessels? Are equipment records complete? 4. Review of previous audits – Have there been any issues cited with MI in the past? 5. Employee interviews – Any other issues not surfaced in the roundtable? 6. Review of follow-up systems in use – What systems are used to keep track of ITPM program inspections? What happens when inspections are not completed in a timely manner? 7. Field tours – Make an informal assessment of MI conditions in the field. Check relief devices in the field versus what is in the files. 8. Review of PSM elements in action (MOC, PSSR, safe work permits, PHA, inspections, etc.) – Track an inspection work order from beginning to end.		

Table 6.2 ***Continued***

Element	PSM Assessment Tasks	Assessment Results	Gap Versus Standard/ Best Practice
[1910.119(k)(1)]. *Hot Work Permit (HWP)*	1. Employees roundtable discussion of PSM issues – Are permits used as required per site procedures? What issues are there with ensuring that all permitting requirements are followed? 2. Best practice sharing by the site and the assessment team – Does the site follow an industry standard HWP format? Does the site exceed the requirements of this standard? Does the site have a routine audit in place to ensure HWP procedure compliance? What sharing does the assessment team have for the site? 3. Records reviews – Review a site cross section of HWPs. 4. Review of previous audits – Have there been any issues cited with HWPs in the past? 5. Employee interviews – Any other issues not surfaced in the roundtable? 6. Review of follow-up systems in use – What systems are used to keep track of HWP compliance issues? 7. Field tours – Check several HWPs in the field. 8. Review of PSM elements in action (MOC, PSSR, safe work permits, PHA, inspections, etc.) – Review past and proposed changes to the HWP procedure.		

Table 6.2 *Continued*

Element	PSM Assessment Tasks	Assessment Results	Gap Versus Standard/ Best Practice
[1910.119(l)(1)]. Management of Change (MOC)	1. Employees roundtable discussion of PSM issues – Are employees involved in MOCs? Are actions from MOCs completed in a timely manner? Are the voices of employees heard during MOCs? Is all training completed in a timely manner? Are the potential hazards from each proposed change reviewed to the appropriate degree prior to implementation? Are MOCs completed for every change on site in PSM areas? 2. Best practice sharing by the site and the assessment team – Is there a routine audit in place to ensure compliance with the site MOC procedure? What type of database is used to track MOCs? Is this system user friendly? Does the current MOC procedure go above and beyond the standard? What sharing does the assessment team have for the site? 3. Records reviews – Has the site been following its MOC procedure? Are MOC actions and training completed in a timely manner? Is there a PSSR for every MOC? 4. Review of previous audits – Have there been any issues cited with MOCs in the past? 5. Employee interviews – Any other issues not surfaced in the roundtable? 6. Review of follow-up systems in use – What systems are used to keep track of MOCs and MOC actions? What happens when actions are not completed in a timely manner? 7. Field tours – Ask "Has an MOC been completed for that recent change?" 8. Review of PSM elements in action (MOC, PSSR, safe work permits, PHA, inspections, etc.) – Pull a completed MOC and verify that all actions have been completed in the field and in documentation as intended. Check on the status of an MOC that is not complete.		

Table 6.2 *Continued*

Element	PSM Assessment Tasks	Assessment Results	Gap Versus Standard/ Best Practice
[1910.119(m)(1)]. Incident Investigation (II)	1. Employees roundtable discussion of PSM issues – Are employees involved in IIs? Are actions from IIs completed in a timely manner? Are the voices of employees heard during IIs? Are investigations conducted for every significant incident? Does the site have trained investigators? 2. Best practice sharing by the site and the assessment team – Is there a routine audit in place to ensure compliance with the site's II procedure? What type of database is used to track IIs? Is this system user friendly? Does the current II procedure go above and beyond the standard? What sharing does the assessment team have for the site? 3. Records reviews – Has the site been following its II procedure? Are II actions completed in a timely manner? Do actions address root causes? 4. Review of previous audits – Have there been any issues cited with IIs in the past? 5. Employee interviews – Any other issues not surfaced in the roundtable? 6. Review of follow-up systems in use – What systems are used to keep track of II actions? What happens when actions are not completed in a timely manner? 7. Field tours – Review the site of the last incident. 8. Review of PSM elements in action (MOC, PSSR, safe work permits, PHA, inspections, etc.) – Attend an II if possible. Track a completed II from start to finish.		

Table 6.2 *Continued*

Element	PSM Assessment Tasks	Assessment Results	Gap Versus Standard/ Best Practice
[1910.119(n)]. Emergency Planning and Response (EPR)	1. Employees roundtable discussion of PSM issues – Are employees involved in EPR? Are drills conducted on a periodic basis? 2. Best practice sharing by the site and the assessment team – Are outside agencies involved in EPR drills? Are outside agencies trained on the possible hazardous release scenarios of the site? Does the site go above and beyond this standard? What sharing does the assessment team have for the site? 3. Records reviews – Is EPR equipment inspected as required? Is all EPR training up to date? Does the level of training meet the standard requirements? Does the EPR drill scope and frequency meet the standard? 4. Review of previous audits – Have there been any issues cited with EPR in the past? 5. Employee interviews – Any other issues not surfaced in the roundtable? 6. Review of follow-up systems in use – What systems are used to keep track of EPR drill critique actions? To keep track of EPR actions for incident investigations? What happens when actions are not completed in a timely manner? 7. Field tours – Inspect EPR equipment. 8. Review of PSM elements in action (MOC, PSSR, safe work permits, PHA, inspections, etc.) – Participate in a routine EPR meeting.		

Table 6.2 ***Continued***

Element	PSM Assessment Tasks	Assessment Results	Gap Versus Standard/ Best Practice
[1910.119(o)(1)]. *Compliance Audits (CA)*	1. Employees roundtable discussion of PSM issues – Are employees involved in CAs? Are results of CAs shared with all employees? Do CAs cover all elements and requirements of the PSM standard? 2. Best practice sharing by the site and the assessment team – Are third-party auditors used to lead CAs? Does the site go above and beyond this standard? What sharing does the assessment team have for the site? 3. Records reviews – Are actions from previous CAs completed in a timely manner? Are CAs completed within the three-year time frame as required? Does each CA cover all requirements of the standard? 4. Review of previous audits – Have there been any issues cited with CAs in the past? 5. Employee interviews – Any other issues not surfaced in the roundtable? 6. Review of follow-up systems in use – What systems are used to keep track of CA actions? What happens when actions are not completed in a timely manner? 7. Field tours – Question facility personnel on their knowledge of past CAs. 8. Review of PSM elements in action (MOC, PSSR, safe work permits, PHA, inspections, etc.) – Follow up on CA actions to verify that they were completed as intended.		

Table 6.2 *Continued*

Element	PSM Assessment Tasks	Assessment Results	Gap Versus Standard/ Best Practice
[1910.119(p)(1)]. *Trade Secrets (TS)*	1. Employees roundtable discussion of PSM issues – Is information withheld from employees due to trade secrets? 2. Best practice sharing by the site and the assessment team – What sharing does the assessment team have for the site? 3. Records reviews – Any issues? 4. Review of previous audits – Have there been any issues cited with TS in the past? 5. Employee interviews – Any other issues not surfaced in the roundtable? 6. Review of follow-up systems in use – Any issues? 7. Field tours – N/A 8. Review of PSM elements in action (MOC, PSSR, safe work permits, PHA, inspections, etc.) – N/A		

The goal of root cause analysis (RCA) is to identify actions we can take to *set up our front-line personnel for success* and eliminate or control factors that set them up for failure.

Recommendations from RCAs can be classified into one of four levels:

- **Level 1:** Address the causal factor (front-line personnel performance gap or equipment performance gap)
- **Level 2:** Address the cause of the specific problem
- **Level 3:** Fix similar, existing problems
- **Level 4:** Correct the business process that creates these problems (the underlying management system performance gap)

To ensure an effective RCA process for your PSM program, Level 4 recommendations must be identified for each significant element or program gap. Level 4 recommendations are aimed at correcting the management system issues that led to a company's PSM performance gaps. The recommendations, if complex (i.e., not simple, straightforward, and obvious), should be tested to ensure that they address these management system issues.

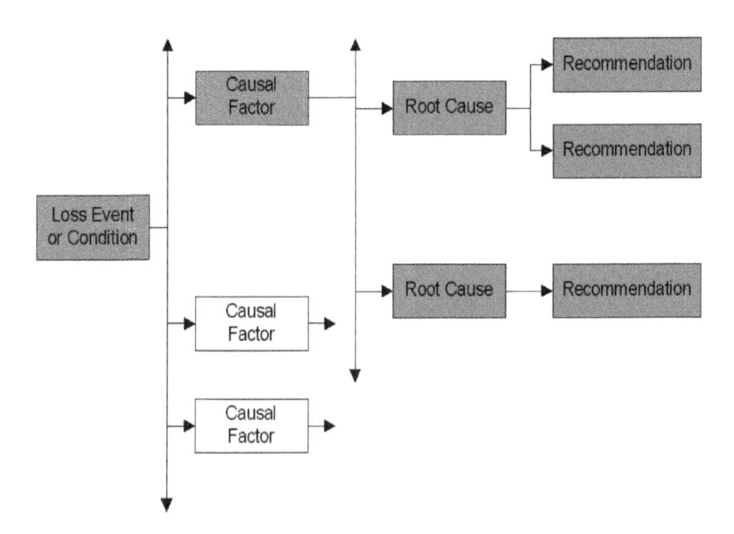

Figure 6.1 Flowchart for Determining Root Causes

6.3 IMPROVING THE PSM PROGRAM

6.3.1 Team Charter

Ahead of selecting the PSM program improvement team, a team charter should be developed. The charter may initially be developed by the project owner (e.g., site manager, area manager, SHE manager) with assistance from the site's PSM coordinator/manager. The charter should document the team's purpose and clearly define roles, responsibilities, and rules. It should also establish procedures for team communication, reporting, and decision-making. The charter may include the following sections:

1. **Purpose.** Describe the purpose of the team and the anticipated outcomes. Much effort should be put into determining the purpose and anticipated outcomes, including seeking input from all disciplines within the organization. The purpose of the team will center around closing key gaps in the PSM program, along with developing recommendations to help close those gaps.
2. **Background.** Summarize the project the team is supporting, state how the team fits within the organizational structure, and identify the project's customers.
3. **Scope.** Describe the scope and objectives and the team's role in addressing these. Define the goals the team must accomplish. Identify the PSM programs, systems, and elements, and the locations, facilities, and units that will or will not be impacted. In this section point the team in the desired direction, but allow them enough leeway to develop the path that will lead to the desired outcomes.
4. **Team composition.** Identify (a) the functional areas required to be represented on the team, (b) the number of members from each, (c) core members versus support or advisory members, (d) full-time or part-time requirements, and (e) the anticipated time/resource commitments needed for the estimated duration of the project.
5. **Membership roles.** Identify the role and responsibilities of each team member. This is a critical part of the team charter. It will (a) help ensure that the work of team members does not overlap and (b) give a clear picture of what each member is expected to accomplish.
6. **Team empowerment.** Define the level of empowerment the team will have. It is very important to let the team know what power they have to "get things done." It is also important to let other functional areas in the facility know this in order to avoid any confusion or conflict.

7. **Team operations.** Describe the team's operational structure, such as its decision-making (consensus or majority) process, plans to establish ground rules or operating guidelines, need for logistical support, etc. The team will be most effective when given the opportunity to establish its own rules. Therefore, this section should be edited by the team once it has been drafted.
8. **Team performance assessment.** Document key areas of performance needed for team success, along with the means of measuring progress. KPIs should include timing, number of resolutions developed for compliance gaps, cost savings, and safety improvements (as well as efficiency gains) as measured by company metrics.
9. **Milestones and schedules.** Include major activities and milestones as part of the project schedule. The team needs to know when delivery is expected for each phase of the project. Milestones are also good checkpoints at which the team can review its progress to date and show management what has been accomplished.
10. **Reporting.** Determine the method and frequency of reporting throughout the project, up to delivery of the final report. With any project, transparency helps ensure that there are no surprises in the final stages of the team's activities.
11. **Signature page.** Each team member should sign the charter to indicate agreement with its contents. This will help ensure their ownership of the project.
12. **Approval.** The project owner approves the team charter, including granting the authorities requested above.

6.3.2 Team Formation

Once the assessment is complete and the charter has been developed, a PSM program improvement team (or one with a similar title) should be formed. The output of the assessment and the value gap analysis (the results of which will be included in the charter) will have significant bearing on the makeup of the team. A cross section of site employees will be required to achieve the success dictated in the charter. Such employees may include, but not be limited to, the following:

- PSM coordinator
- Maintenance planner
- Reliability/inspection specialist
- Mechanical, electrical, and instrumentation technicians
- Operators
- Area managers
- Engineers

- Supervisors
- Safety, health, and environmental personnel
- Third-party representatives with PSM expertise

A wide-ranging scope encompassing most of the PSM elements will require a very large team or a team that includes several specialist subteams. A more targeted scope, such as one dealing with mechanical integrity, will require a smaller, more focused team.

A team leader should be designated. Excellent candidates include PSM coordinators/managers, engineering/operations/maintenance managers, or other proven "leaders" at the site.

Choosing the right team for the task

General characteristics of effective process improvement teams include the following:

- **Relevant expertise.** Whether a new system is being created or an existing one modified, the people responsible must have enough experience and judgment to provide meaningful input. The experience of the team members provides credibility and promotes confidence in the overall team and effort.
- **Stakeholder representation.** Every process has owners, customers, and suppliers, and wherever possible all of these stakeholder groups should be represented on the PSM program improvement team. This representation promotes transparency, trust, and confidence in the overall team and effort.
- **Sufficient authority.** PSM program improvement often requires working across organizational boundaries, and in some companies this requires specific authority. In addition, to be effective the team and its leader should be adequately empowered to undertake the task at hand without needing multiple approvals every step of the way.
- **Clearcut mission.** As with other team structures, the team needs a clearly articulated understanding of its charter, responsibilities, and limits.

In considering team composition, you will probably find that many candidates have emerged through your PSM work to date. Members of the PSM assessment team, because of their exposure to the initiative, are logical leaders for program improvement teams and subteams; their continuing involvement also helps reinforce consistency.

6.3.3 Project Initiation

With the charter developed and the team formed, you can now begin to develop improvements to the PSM programs that have been selected and prioritized using the value gap analysis and the PSM assessment. The initial team meeting should cover topics such as:

- charter review;
- further scope development;
- setting of ground rules;
- discussion led by the team leader of management's expectations (e.g., active involvement, time allocation, leadership, results driven activity);
- required team training (PSM, teamwork, etc.); and
- desired outcomes, including the final work product.

6.4 DEVELOPING THE SOLUTION FOR AN ELEMENT OR SYSTEM

6.4.1 Team Output

Once the scope has been set, the team will act on the results of the PSM assessment. The team will determine how to manage the various sections of the scope, which may include developing subteams for the various elements.

The team members will utilize recommendations from the PSM assessment or will be requested to develop their own recommendations based on their knowledge of and experience with facility systems.

Teams will be most effective when functioning as working teams, as opposed to teams that assign work to be done. Working teams are those that actually act on the recommendations they develop and help ensure that these actions are implemented as desired by the team (or something close to what the team wanted). Working teams work alongside area teams to provide the needed expertise and assistance during the implementation stage. The teams will normally require assistance to accomplish work, and this will come from personnel from the area teams. The teams should stay intact until the work is complete. Prioritized actions plans should be developed and tracked along the way. The action plans will be assigned ownership with due dates. A tracking database should be utilized for all team actions.

The teams' recommendations must be feasible and provide gap closure identified by the assessment. Acceptable recommendations must:

- address options for reducing the frequency and/or consequences of one or more causal factors or root causes;

- clearly state the intended action;
- be practical, feasible, and achievable;
- not pose other unacceptable risks;
- be based on conclusions from data collected during the assessment/gap analysis; and
- provide a general objective to be accomplished, followed by a specific example of how it *could be* successfully accomplished.

6.4.2 Management Responsibility for Recommendations

Management is key to the success of the outcome of this project. To that end, management must be willing to accept responsibility for the gaps identified in the assessment/gap analysis, and be willing to accept responsibility for assessing and managing the team's recommendations. As far as recommendations go, management responsibilities include:

- Review recommendations to evaluate feasibility, practicality, and effectiveness
- Establish schedules for implementing accepted recommendations
- Assign individuals the responsibility of implementing accepted recommendations
- Evaluate recommendations as management of change items (when applicable)
- Provide affected personnel with the necessary training related to recommendations
- Document resolutions
- Track recommendations to completion
- Look for opportunities to reduce risk in other systems

6.5 MONITORING IMPROVEMENT OF AN ELEMENT OR SYSTEM

6.5.1 Check Output Versus Established Criteria

As you near completion of each PSM program improvement project, it is useful to check results against the criteria you established in the charter. You may find it helpful to review the plan at intervals throughout the life of the project. Similarly, you will want to ensure that the actions you develop address the priorities you identified as part of the planning process.

By checking against these criteria periodically, you can make any necessary refinements or adjustments before undertaking full implementation. You should also consider a final, global validation of all the system components to make sure they integrate properly.

6.5.2 Measure and Monitor Installation

By the time you have successfully tested and refined your new PSM systems and installation is underway, you and your team have passed a significant milestone. Congratulations are in order – but there is still more to be done. The good news is that at this point most of the effort shifts from the PSM team to your company's facilities. As a result, your role will change from directing a project to supporting and monitoring its local implementation. During the PSM installation phase, activity will be spread widely throughout your company. Whether you have chosen a centralized or a decentralized approach for PSM design and development, installation must occur at the facility level. It is important to keep in mind that individual facilities will be the ultimate owners and operators of the PSM systems at their sites, and they must assume responsibility during installation. However, there is still a need during this stage to maintain an overview of progress and ensure that you are meeting the objectives of the original PSM plan. Measuring and monitoring PSM installation helps ensure consistency and quality control; in addition, effective monitoring helps keep local expenditures and schedules on track. In effect, these activities protect the investment that you, your team, and your company have made during the course of your work so far, and for this reason they warrant continuing attention.

6.5.3 Metrics

Metrics are required to help monitor the performance of your PSM program. Some key resources for information on PSM metrics needed to monitor your program are listed below:

- The CCPS metrics Web site at www.aiche.org/ccps/search/metrics, which provides a wealth of information, including brochures, presentations, and Webinars on this subject
- The 40-plus-page CCPS brochure entitled *Process Safety Leading and Lagging Indicators . . . You Don't Improve What You Don't Measure* (Ref. 6.1)
- The CCPS book *Guidelines for Process Safety Metrics* (Ref. 6.2), which provides basic information on process safety performance indicators, including a comprehensive list of metrics for measuring performance and examples on how they can be successfully applied both short term and long term
- Similarly, ANSI/API RP 754 (Ref. 6.3), which provides possible metrics, grouped into four tiers

6.5.4 Customer Feedback and Follow-up

For a management system to work effectively, users must accept it and contribute to its continuous improvement. For this reason you should consider obtaining "customer" feedback on a PSM system after it has been installed and operating for a short time. Such feedback, from anyone who interacts with the system by providing input, using information, or receiving reports, can greatly enhance the effectiveness of the new PSM system. System users may have important suggestions on how to improve or streamline a system, or they may not clearly understand its importance or the part they play in its implementation. At the same time, soliciting – and acting on – user feedback helps underscore the collaborative nature of PSM and the fact that its success depends on the user.

Interviews conducted during onsite progress reviews provide one source of user feedback. However, it is likely that you will speak with only a small number of the people who interact with the PSM system. To acquire broader input, you should consider a formal user survey. Written surveys request feedback on the PSM system, asking about its clarity, ease of use, quality of training received on the system, perceived barriers to its effective use, perceived effectiveness in meeting objectives, and suggested modifications and improvements. These types of surveys typically mix open-ended and "yes/no" questions and are distributed to all system users. They tend to be simple and brief to encourage participation.

The results of a user survey should not be used as a "pass/fail" test of the system's success. Instead, consider the results as a way of helping to fine-tune the system. To make the most of this exercise, retain quantitative results for comparison against future surveys to help measure progress in continuous improvement.

If you decide to conduct a PSM user survey, keep in mind that it is essential to provide timely feedback to respondents on the survey results and the action plans or system changes (if any) that result. If users see that their input has value and impact, they are more likely to develop the sense of ownership that will help determine the long-term success of the PSM system.

The installation of a PSM system does not complete the system development; it begins the process of continuous improvement. All management systems can be improved over time. Even if a "perfect" system could be designed, changing business conditions, personnel, organizational structure, regulations, and technical knowledge require that PSM management systems evolve over time. PSM management systems should be regularly assessed to identify improvement opportunities. Deficiencies discovered during PSM audits may highlight a need for system improvement. PSM auditing is discussed in detail in the CCPS book *Guidelines for Auditing Process Safety Management Systems, 2nd Edition* (Ref. 6.4).

6.6 REFERENCES

6.1 Center for Chemical Process Safety of the American Institute of Chemical Engineers, *Process Safety Leading and Lagging Metrics . . . You Don't Improve What You Don't Measure,* New York, New York, January 2011, www.aiche.org/sites/default/files/docs/pages/CCPS_ProcessSafety_Lagging_2011_2-24.pdf.

6.2 Center for Chemical Process Safety of the American Institute of Chemical Engineers, *Guidelines for Process Safety Metrics*, John Wiley & Sons, Inc., Hoboken, New Jersey, 2009.

6.3 ANSI/API Recommended Practice 754, *Process Safety Performance Indicators for the Refining and Petrochemical Industries*, American Petroleum Institute, Washington D.C., April 2010, www.publications.api.org/.

6.4 Center for Chemical Process Safety of the American Institute of Chemical Engineers (AIChE, *Guidelines for Auditing Process Safety Management Systems, 2nd Edition*, John Wiley & Sons, Inc., Hoboken, New Jersey, 2011.

7

INTEGRATING PSM/HSE WITH A BUSINESS MANAGEMENT SYSTEM

What is a business management system (BMS)? Here are a few representative definitions:

- "A set of policies, practices, procedures, and processes used in developing and deploying strategies, their execution, and all associated management activity." (www.BusinessDictionary.com)
- "A set of interrelated or interacting elements to establish policy and objectives and to achieve those objectives." (ISO 9000:2000)
- "The structure, processes and resources needed to establish an organisation's policy and objectives and to achieve those objectives." (Chartered Quality Institute)
- "A set of tools for strategic planning and tactical implementation of policies, practices, guidelines, processes and procedures that are used in the development, deployment and execution of business plans and strategies and all associated management activities." (www.MyManagementGuide.com)
- "A set of tools for strategic planning and tactical implementation of policies, practices, guidelines, processes and procedures that are used in the development, deployment and execution of business plans and strategies and all associated management activities." (The Law Dictionary)

Many companies have instituted BMSs as a vehicle to improve the performance, consistency, efficiency, and other aspects of their operations. Most of these companies who also have a PSM or HSE system have recognized (1) that there are many interactions between the BMS and the PSM/HSE system that have potentially negative impacts on the PSM/HSE system and (2) the importance of integrating (or better integrating) the PSM/HSE system with their BMS.

This chapter (1) focuses on possible BMS and PSM/HSE system interactions with potentially negative impacts, (2) discusses how to prevent or minimize them, and (3) provides general guidance on the important subject of integrating PSM/HSE with the BMS.

7.1 VALUES AND POLICY INTERFACES/CONFLICTS WITH BUSINESS ENTERPRISE

Since one of the primary objectives of any business in a free enterprise system is to achieve profits in order to stay in business, there is always a natural tendency for the organizations and the individuals in them to focus on this objective, potentially at the expense of their expressed values and policies implemented to achieve objectives such as striving for zero injuries or fatalities, minimizing process safety incidents, protecting the neighbors and the environment, etc. In other words, companies may focus too much on profits at the expense of safety.

Most businesses (or companies) have missions and/or visions that are designed to focus and guide them. These mission/vision statements should include the value placed on employees, contractors, neighbors, and the environment, especially for companies that handle or produce hazardous materials in the course of business. These companies must place these values above profits and understand that ultimately these values enable a profitable business. On the other hand, placing profits ahead of these values will eventually cause the business to fail. See Chapter 3 in this book, Chapter 3 in the RBPS book, and the other materials referenced there for more information on process safety culture and its importance to process safety performance.

7.2 TYPES OF BMS ACTIVITIES

Many types of systems exist and activities performed within the typical BMS, and most if not all of these can impact the PSM system and PSM/HSE performance. These impacts will be negative if the activity or interfaces with the PSM/HSE systems are not managed properly. Following is a list of such activities, possible behaviors, their potential PSM/HSE impacts, and things to consider to avoid or minimize negative impacts:

- **Business model/profit and loss goals.** A company's business model, and particularly the way in which profit and loss goals are set, can potentially drive the internal organizations toward negative behaviors with negative PSM/HSE impacts. Examples of these and measures to prevent or reduce the behaviors/impacts are summarized in Table 7.1.
- **Budgeting.** In a different way but with similar results as from its business model, the company/business/site budgeting process can drive negative behaviors and PSM/HSE impacts; these can be prevented or reduced with similar measures. See Table 7.1.

- **Capital expenditure (Capex).** Capex processes and activities are typically developed by corporate/site engineering groups and driven by those groups and individual project teams. Capex can drive behaviors such as minimizing project costs, maintaining the project schedule, and performing inadequate MOC reviews, all of which can weaken the PSM system in one way or another. See Table 7.2 for additional information on these, their impacts, and preventive measures.

Table 7.1 Business Model and Budgeting Behaviors, Impacts, and Preventive Measures

Possible Behaviors	Potential Negative Impacts	Preventive Measures
Focusing on profits much more than on PSM/HSE performance	A poor process safety culture	• Developing and maintaining a sound process safety culture (see Chapter 5) • Reminding stakeholders of mission/vision/values and the need for process safety to enable business value
Unwillingness to add new PSM elements/activities	Lack of continuous improvement in PSM/HSE	• Ensuring that the addition of new PSM elements/activities is periodically considered (e.g., by instituting a management review and continuous improvement element [see Chapter 5]) • Implementing and institutionalizing process safety metrics (see Chapter 5) to identify and drive opportunities, including new PSM elements or activities
Delaying large, ongoing expenditures related to PSM/HSE in favor of short-term profits (e.g., repeatedly delaying turnarounds, scheduled preventive maintenance, painting)	Increased loss of containment incidents and equipment breakdowns	• Implementing and institutionalizing process safety metrics (see Chapter 5), including ones that ensure that routine maintenance, required inspections/tests, and follow-up repairs are performed as required by the PSM system • Developing and maintaining a sound process safety culture (see Chapter 5), including educating management (decision-makers) on process safety
Maintaining or reducing "headcount" when there may not be enough people to sustain or achieve planned improvements in the PSM system	All of the above	All of the above

Table 7.2 Capex Behaviors, Impacts, and Preventive Measures

Possible Behaviors	Potential Negative Impacts	Preventive Measures
Minimizing project costs	• Installation of the cheapest (rather than the most cost-effective) equipment, leading to increased breakdowns/incidents • Failure to provide full, complete project documentation, leading to initial and ongoing inaccuracies in the process safety information	• Developing and maintaining a sound process safety culture (see Chapter 5) • Implementing a rigorous asset integrity and reliability element (see Chapter 12 of the RBPS book)
Maintaining the project schedule	• Performance of PHAs with inadequate design and process safety information • Inadequate project installation reviews ("punch lists") and PSSRs, and/or inadequate equipment/system function checks, leading to poor operation and increased breakdowns/incidents	• Developing and maintaining a sound process safety culture (see Chapter 5) • Developing a project management work process that establishes expectations for PSM elements (PSI, PHA, operating procedures, training, MOC, PSSR, MI, etc.)
Performing inadequate MOC reviews	• Large projects and/or changes not adequately reviewed against PSM system requirements, including PHAs, leading to inadequate design and increased breakdowns/incidents	• Implementing and maintaining a rigorous MOC element (see Chapter 15 of the RBPS book) • Developing a project management work process that establishes expectations for PSM elements (PSI, PHA, operating procedures, training, MOC, PSSR, MI, etc.)

- **Quality.** Although quality (e.g., ISO 9001) and PSM/HSE systems can be successfully integrated, quality systems are sometimes segregated from and in conflict with PSM/HSE systems, potentially driving organizations toward negative behaviors and PSM/HSE impacts. Examples of these and measures to prevent or reduce them are summarized in Table 7.3.
- **Security of information.** Information security-related activities within an organization's BMS are generally not a problem. However, if they go to the extreme of promoting behaviors where PSI or other PSM information

is not fully or easily shared with employees or contractors (who need the information for work they are performing) due to "confidentiality" or "trade secrets" concerns, then the pendulum has swung too far. Open access to such information should be provided, with training and/or confidentiality agreements used to ensure adequate control of sensitive information.

- **Information technology (IT)/information management systems (IMS).** As discussed in Chapter 4, modern PSM systems typically require some IT/IMS tools for successful operation. Conversely, IT and IMS groups sometimes have excessive autonomy and/or a lack of customer focus, which can lead to the "tail wagging the dog." In other words, this can result in an environment that makes it difficult, if not impossible, to obtain, implement, and support all of the computer-based tools that are needed (or at least cause significant delays and extra efforts). Therefore, it is vital for organizations to develop and maintain a sound safety culture (see Chapter 5) and, in particular, to continuously nurture the essential feature of establishing safety as a core value.

Table 7.3 Quality Behaviors, Impacts, and Preventive Measures

Possible Behaviors	Potential Negative Impacts	Preventive Measures
Focusing on and/or discussing product quality as much or more than PSM/HSE performance	A poor process safety culture	• Developing and maintaining a sound process safety culture (see Chapter 5) that drives an understanding of how quality relates to process safety risk
Minimizing details in procedure/policy documents in order to "say what we do and do what we say"	Increased operating issues, loss of containment incidents, and equipment breakdowns	• Developing and maintaining a conduct of operations and operational discipline element (see Chapter 5)
Overemphasizing "quality critical" parameters, instruments, etc. (at the expense of PSM/HSE equivalents)	"Dilution" of attention to PSM/HSE critical parameters, instruments, etc., leading to increased process safety incidents and equipment breakdowns	• Developing and maintaining robust process knowledge documentation and operating procedures (see Chapters 8 and 10 of the RBPS book) • Implementing and maintaining a rigorous asset integrity and reliability element (see Chapter 12 of the RBPS book)

- **Human resources (HR).** In some organizations, the HR function controls hiring, training, promotions and raises, discipline, and retirements with little or no consideration/appreciation for the potential negative impacts on the PSM/HSE system and performance. Examples of potential personnel-related BMS behaviors, negative impacts, and preventive measures are provided in Table 7.4.
- **Hiring and professional development.** The HR BMS or other related ones typically address hiring (of wage roll, salary, technical, and management personnel) and professional development. Hiring and professional development activities can drive negative behaviors and PSM/HSE impacts. See Table 7.4 for examples of these as well as preventive measures.
- **Competency.** Similarly, competency-related activities within the BMS (or lack thereof) can drive negative behaviors and negatively impact PSM/HSE. Examples of these and measures to prevent or reduce the behaviors/impacts are summarized in Table 7.4.
- **Apprenticeship and job rotation.** A final area of personnel-related BMS activities deals with apprenticeship and job rotation. In many cases, the "rules" in this area are based on longstanding union contracts, which the organization may or may not have challenged in the past. In any event, relying too much on apprenticeship programs (at the expense of individual competency) or excessively limiting job rotation can have significant negative PSM/HSE impacts. See Table 7.4 for examples of these as well as preventive measures.
- **Sustainable development.** "Sustainability" is a common goal of petrochemical manufacturing organizations, and it generally focuses on minimizing waste and pollution from manufacturing processes to ensure the most efficient operation over the long term. However, activities should not be given a "pass" from meeting the PSM system requirements simply because they will make a process or operation more sustainable. An example is when many facilities/processes started collecting process vents in common headers to flares and similar devices, using blowers, vacuum pumps, etc. In many cases, these "environmental improvement" projects did not receive adequate MOC reviews and PHAs, and subsequently explosions occurred within the collection systems (see the CCPS book *Safe Design and Operation of Process Vents and Emission Control Systems* [Ref. 7.1] for more information on this important subject). Therefore, organizations must ensure that PSM requirements are followed by everyone all the time, particularly as they relate to MOC reviews for new/modified facilities.

Table 7.4 Personnel-related Behaviors, Impacts, and Preventive Measures

Possible Behaviors	Potential Negative Impacts	Preventive Measures
Maintaining or reducing "headcount" when there may not be enough people to sustain or achieve planned improvements in the PSM system (for example, by not hiring additional people or replacing retirees, or through terminations or retirement incentives)	• A poor process safety culture • Inadequate personnel available, leading to increased operational issues, loss of containment incidents, and equipment breakdowns	• Developing and maintaining a sound process safety culture (see Chapter 5), including a management of organizational change system
Implementing an operator/technician training system that focuses on aspects such as time in grade, tests, or skills demos with inadequate focus on operating/maintenance procedures and individual competency	• Inadequate training and/or verification of understanding, leading to increased operating issues, loss of containment incidents, and equipment breakdowns	• Implementing and maintaining a rigorous training and performance assurance element (see Chapter 14 of the RBPS book)
Promoting based on getting results and/or how well liked a person is, with little or no consideration of their PSM/HSE performance and priority	• A poor process safety culture • Personnel in supervisory positions who lack PSM/HSE competency, leading to increased operating issues, loss of containment incidents, and equipment breakdowns	• Developing and maintaining a sound process safety culture (see Chapter 5) • Implementing a process safety competency element (see Chapter 5)
Discouraging or limiting the use of disciplinary action for poor PSM/HSE performance or violating cardinal rules	• A poor process safety culture • Ongoing poor PSM/HSE performance, leading to increased process safety incidents, injuries, and near misses	• Developing and maintaining a sound process safety culture (see Chapter 5) • Implementing a conduct of operations and operational discipline element (see Chapter 5)

Table 7.4 *Continued*

Possible Behaviors	Potential Negative Impacts	Preventive Measures
Failure to encourage and support professional development	• Weakness in areas such as training of engineers, knowledge of and compliance with standards, and overall process safety competency, leading to a weaker PSM system and increased process safety incidents	• Implementing and maintaining a rigorous training and performance assurance element (see Chapter 14 of the RBPS book) • Implementing a compliance with standards element (see Chapter 5) • Implementing a process safety competency element (see Chapter 5)
Lack of support for process or corporate technology stewards and technology manuals/documentation	• Poor process safety competency and lack of a learning organization, leading to increased breakdowns, operational upsets, and incidents	• Implementing a process safety competency element (see Chapter 5)
Placing personnel in jobs/roles based solely on their progression through an apprenticeship program	• Inadequate training and/or verification of understanding, leading to increased operating issues, loss of containment incidents, and equipment breakdowns	• Implementing and maintaining a rigorous training and performance assurance element (see Chapter 14 of the RBPS book)
Overly restrictive job rotation (e.g., between field and board or between processes)	• Lack of in-depth knowledge of the process and its operation, leading to increased operating issues, loss of containment incidents, and equipment breakdowns	• Implementing and maintaining a rigorous training and performance assurance element (see Chapter 14 of the RBPS book)

- **Change management.** In some organizations, the business-driven change management philosophy can be at odds with the PSM MOC requirements. For instance, companies may pride themselves on being able to rapidly make organizational changes in order to attain a competitive advantage when market conditions change or new products are available. Such changes can have negative PSM/HSE impacts, particularly if personnel with key PSM/HSE responsibilities are moved or overloaded with new, additional responsibilities. These negative impacts can be avoided or reduced by (1) developing and maintaining a sound process safety culture (see Chapter 5) and (2) ensuring that PSM system procedures and workflows (as applicable) are adequately documented and followed (see Chapter 4).

- **Asset management.** Many corporate/site BMSs include an asset management component and activities. This is not an issue unless it includes principles or activities that may conflict with the requirements of the organization's MI program (or asset integrity and reliability program if the organization has adopted the broader RBPS element [see Chapter 12 of the RBPS book] or a similar element including a reliability focus] in its PSM system. As for other BMS conflicts with PSM or HSE requirements, the PSM/HSE requirements should take precedence, and will if a sound process safety culture (see Chapter 5) is in place.

  Note: While requirements should take precedence, not all PSM/HSE issues are requirements. Some are good practice and some are available options. Choosing a PSM/HSE option that best fits the BMS and still meets the requirements is desirable.

- **Purchasing goods and services.** Another common BMS component that may overlap or conflict with the PSM MI (or similar) element is the purchasing of goods and services. Examples of potential purchasing BMS behaviors, negative impacts, and preventive measures are provided in Table 7.5.

- **Global supply chain.** Similarly, the global scope of the supply chain, especially for larger companies, can result in or encourage some of the same behaviors as the purchasing BMS, as well as introduce other potential behaviors and negative PSM/HSE impacts. See Table 7.5 for examples of such behaviors, negative impacts, and preventive measures.

- **Global customer service.** Businesses that deal with customers on a global basis may exhibit behaviors similar to those for the purchasing or supply chain BMS and introduce other behaviors with potential negative PSM/HSE impacts. Examples of potential customer service BMS behaviors, negative impacts, and preventive measures are included in Table 7.5.

- **Production or manufacturing scheduling.** An organization's BMS approach to production or manufacturing scheduling may drive behaviors with potential negative PSM/HSE impacts. Table 7.6 provides examples of potential scheduling BMS behaviors, negative impacts, and preventive measures.

Table 7.5 Purchasing, Supply Chain, and Customer Service Behaviors, Impacts, and Preventive Measures

Possible Behaviors	Potential Negative Impacts	Preventive Measures
Purchasing from the "lowest bidder" without consideration of the PSM/HSE impact	• Installation of cheapest (rather than most cost effective) equipment, leading to increased incidents/breakdowns	• Developing and maintaining a sound process safety culture (see Chapter 5) • Implementing a rigorous asset integrity and reliability element (see Chapter 12 of the RBPS book)
Substituting a cheaper or different item without ensuring that it is an acceptable replacement	• Installation of equipment that is not suitable and fails, leading to increased incidents/breakdowns	• Implementing a rigorous asset integrity and reliability element (see Chapter 12 of the RBPS book) • Implementing a rigorous management of change element (see Chapter 15 of the RBPS book)
Choosing a contractor to perform work on the site that is not on the approved contractor list or lack of oversight of contractors performing work that could impact process safety	• Increased onsite contractor injuries • Improperly performed maintenance, inspection, or construction work, leading to increased incidents/breakdowns	• Implementing a rigorous contractor management element (see Chapter 13 of the RBPS book)
Spot purchasing of a raw material that meets most but not all established specifications	• Increased corrosion, fouling, unexpected reaction, etc., leading to increased incidents and equipment breakdowns	• Implementing a rigorous process knowledge management element (see Chapter 8 of the RBPS book) that includes applicable upper/lower limits on raw material specifications
Transferring a raw material, intermediate, or product to a facility or customer that has not handled it before	• Increased corrosion, fouling, unexpected reaction, etc., leading to increased incidents and equipment breakdowns • Inadequate review of potential safety issues, leading to increased process safety incidents	• Implementing a rigorous process knowledge management element (see Chapter 8 of the RBPS book) • Implementing a rigorous management of change element (see Chapter 15 of the RBPS book)

Table 7.5 *Continued*

Possible Behaviors	Potential Negative Impacts	Preventive Measures
Asking a facility to produce a new product that has not been thoroughly researched on a lab/pilot scale	• Increased corrosion, fouling, unexpected reaction, etc., leading to increased incidents and equipment breakdowns • Inadequate review of potential safety issues, leading to increased process safety incidents	• Developing and maintaining a sound process safety culture (see Chapter 5) • Implementing a rigorous process knowledge management element (see Chapter 8 of the RBPS book) • Implementing a rigorous management of change element (see Chapter 15 of the RBPS book)
Asking a facility to make a new product grade or blend requiring operation outside of established operating limits	• Increased corrosion, fouling, unexpected reaction, etc., leading to increased incidents and equipment breakdowns	• Implementing a rigorous process knowledge management element (see Chapter 8 of the RBPS book) • Implementing a rigorous management of change element (see Chapter 15 of the RBPS book)

- **Mergers and acquisitions.** Mergers and acquisitions are a part of doing business and an ongoing challenge in the global industry. However, a company's approach to evaluating, implementing, and integrating new facilities or businesses can result in a wide variety of negative PSM/HSE impacts. The CCPS book *Guidelines for Process Safety Acquisition Evaluation and Post Merger Integration* (Ref. 7.2) discusses the potential pitfalls and how they can be avoided or minimized. The book's appendices and accompanying electronic files also include (1) extensive checklists of process safety issues that should be investigated or addressed, (2) a draft of a possible integration plan, and (3) a draft integration budgeting tool.
- **Contract manufacturing.** Many company business models include the use of contract manufacturing, on either a temporary or permanent basis, typically to either (1) produce relatively small volumes of new or low-volume products or intermediates or (2) supplement internal production. In these cases, the ongoing challenges include ensuring that (1) decisions on contract manufacturing operations include consideration of the potential PSM/HSE impact by knowledgeable personnel, (2) a conscious, informed decision is made regarding whether the contract manufacturer's PSM system is adequate or whether they need to meet some or all of the parent company's PSM requirements, and (3) adequate systems are in place to ensure that the established requirements are met. If these challenges are not addressed, a wide variety of negative PSM or HSE

impacts (beyond the scope of this book) could occur. See the CCPS book *Guidelines for Process Safety in Outsourced Manufacturing Operations* (Ref. 7.3) for more information.

- **Business continuity planning.** The BMS of many companies includes business continuity planning; i.e., planning for how to respond to any events that could impact operations (e.g., supply chain interruption, loss of or damage to critical infrastructure). The potential negative PSM/HSE impact from such planning arises when the plans would or could result in temporarily bypassing PSM system requirements. Due to the wide variety of actions that might be taken, we will not attempt to provide examples in this book. However, it should be noted that preplanning, along with a rigorous MOC system and backed by a sound safety culture, should help mitigate the potential negative impacts from most likely business continuity plans/actions.

Table 7.6 Scheduling Behaviors, Impacts, and Preventive Measures

Possible Behaviors	Potential Negative Impacts	Preventive Measures
Delaying a turnaround or maintenance outage past established intervals for inspections/tests of process equipment	• Failure of process equipment or piping leading to increased loss of containment incidents	• Developing and maintaining a sound process safety culture (see Chapter 5) • Implementing a rigorous asset integrity and reliability element (see Chapter 12 of the RBPS book)
Pressuring to shorten planned turnarounds and maintenance outages, and/or to start back up more quickly than prescribed by the operating procedures	• Failure to (1) perform inspections/tests at prescribed intervals, (2) repair or replace disabled/bypassed controls, backup equipment, etc. (or performing work while running (i.e., higher risk) or (3) follow operating procedures leading to increased operational issues, incidents, and breakdowns	• Developing and maintaining a sound process safety culture (see Chapter 5) • Implementing a rigorous asset integrity and reliability element (see Chapter 12 of the RBPS book)
Operating a process with clamps or other temporary leak repairs for excessive periods	• Failure of clamps or other leak repairs leading to increased loss of containment incidents • Root cause not addressed, so additional failures in the same system likely	• Developing and maintaining a sound process safety culture (see Chapter 5) • Implementing a rigorous asset integrity and reliability element (see Chapter 12 of the RBPS book) and/or management of change element (see Chapter 15 of the RBPS book) that addresses and controls clamps and similar leak repairs

7.3 COMPANY AND REGIONAL POLITICS

Even when BMS and PSM/HSE conflicts are identified and possible solutions developed, the resolution must generally be accomplished within the company or through regional "politics." In other words, there may be key people who have to be persuaded to change, channels that must be gone through, a "that's not the way we do things around here" attitude that must be changed, etc. It is not within the scope of this book to attempt to describe the variety of politics that exist within organizations or how to work through them. However, it is a factor that the reader should be aware of and prepared to address on a case-by-case basis.

7.4 WORKFLOWS/PROCESSES OF EXISTING BMS

Just as a well-designed PSM system has workflows and work processes (see Chapter 4), a well-designed BMS does as well. Therefore, those working to integrate PSM/HSE systems with BMS should compare these where they interact (e.g., in the areas noted in Section 7.2) to determine what changes may be required (in either or both systems) to minimize conflicts and negative impacts and improve performance of the systems.

7.5 PLANNED CHANGES TO EXISTING BMS

When working to integrate PSM/HSE systems with the existing BMS, it is worthwhile to be aware of any near-term changes that are planned for the BMS. By doing this, changes can be made, as necessary, to the PSM/HSE system and/or the BMS implementation plans in a way that maximizes synergy and improves the integration between the two systems.

7.6 INTERFACES WITH EXISTING BMS

Section 7.2 discusses a number of systems or activities where interfaces between the PSM/HSE system and the BMS are likely to exist. However, these interfaces may not all exist and there may be additional ones. Therefore, it is important to review both existing and planned systems closely to ensure that all applicable interfaces have been catalogued and can then be addressed in the integration effort.

7.7 RESOLVING BMS CONFLICTS

Once all of the potential conflicts have been identified, the effort to integrate (or better integrate) the PSM/HSE system with the BMS should (1) identify actual and potential conflicts and (2) develop and implement plans to resolve each, using the preventive measures provided in Section 7.2 as a starting point.

7.8 REFERENCES

7.1 Center for Chemical Process Safety of the American Institute of Chemical Engineers, *Safe Design and Operation of Process Vents and Emission Control Systems*, John Wiley & Sons, Inc., Hoboken, New Jersey, 2006.

7.2 Center for Chemical Process Safety of the American Institute of Chemical Engineers, *Guidelines for Process Safety Acquisition Evaluation and Post Merger Integration*, John Wiley & Sons, Inc., Hoboken, New Jersey, 2010.

7.3 Center for Chemical Process Safety of the American Institute of Chemical Engineers, *Guidelines for Process Safety in Outsourced Manufacturing Operations*, John Wiley & Sons, Inc., Hoboken, New Jersey, 2000.

8

MANAGING FUTURE PROCESS SAFETY PERFORMANCE

If a company or site treats PSM implementation as a project, then just like the process equipment installed in a capital project, the PSM system is likely to "corrode" if it is not properly monitored and maintained. Therefore, it is vital to have a plan and activities in place to manage future process safety performance.

This chapter provides guidance on how to ensure that a robust PSM system is in place, avoid past PSM system failure modes, and be aware of early warning signs of process safety failures.

8.1 ENSURE A ROBUST PSM SYSTEM

The *Guidelines for Risk Based Process Safety* (Ref. 8.1) state that "safe operation and maintenance of facilities that manufacture, store, or otherwise use hazardous chemicals requires robust process safety management systems." In this context, a "robust" system might be defined as "a system which has the ability to resist unintended change without losing its initial stable configuration" (adapted from Wikipedia). A robust PSM system is also fault tolerant, meaning it can "continue operating properly in the event of the failure of (or one or more faults within) some of its components" (also adapted from Wikipedia). Furthermore, a robust PSM system is resilient; that is, it can deal with changes over time. Finally, a robust PSM system needs to be sustainable; that is, one that will stand the tests of time.

So, whether a site or company is implementing a new PSM system, adding elements, or enhancing existing elements, the goal should be to create an increasingly robust, fitness-for-purpose and sustainable PSM system.

8.1.1 Critical Success Factors for a Robust PSM System

Following are some of the critical success factors for implementing and sustaining a robust PSM system:

- **Willingness to improve and consider changes.** In order to pursue a goal of zero process safety incidents (including near misses), an organization has to be willing to change. Although the way you did some things may have served you well or at least adequately, to achieve optimum performance, change – and sometimes radical change – may be required in some activities and approaches.

- **Commitment to fitness-for-purpose.** A recurring, underlying theme in *Guidelines for Risk Based Process Safety* is that PSM elements and the associated management systems must possess fitness-for-purpose (see Table 8.1). And no matter how "fit" the systems are, they will not be robust if the individuals responsible for their execution do not perform to a high standard.
- **Commitment to process safety culture and operational discipline.** As discussed in Chapters 3 and 5 of this book and at length in *Guidelines for Risk Based Process Safety* and *Conduct of Operations and Operational Discipline for Improving Process Safety in Industry* (Ref. 8.2), the underlying culture of the organization and the individual and organizational discipline to perform each task the right way every time are vital to implementing any PSM system change. They are equally vital to making the system robust and sustainable.

  Note: There is no substitute for providing the management leadership required to create and sustain the essential features of a good safety culture (see Table 8.2), particularly the first three features.
- **Create an effective learning organization.** Many books, papers, and presentations have been developed in the last few years on the importance of creating PSM-related learning organizations (Refs. 8.3 through 8.9). Important factors in creating an effective learning organization include:
  - applying root cause thinking to all of the "deviations" that occur in the PSM system (not just to incidents) so that any recurring performance issues can be identified and corrected;
  - maintaining an effective corrective action process – one that enables corrective actions generated by PSM system elements to be resolved and properly implemented in a timely fashion;
  - performing high-quality incident investigations, applying the time and resources necessary to identify root causes;
  - monitoring proper process safety metrics and acting on negative trends (see the discussion in Chapter 5 on measurement and metrics);
  - performing discerning audits that not only look at the PSM "paperwork," but also focus on actual PSM performance and any underlying systemic issues; and
  - using effective management reviews to periodically examine past activities, current performance, and trends, and actively discussing areas for continuous improvement in each PSM system element (see the discussion in Chapter 5 on management review and continuous improvement).

Table 8.1 Fitness-for-purpose Summary

Principle: Management systems should be the simplest they can be while still possessing fitness-for purpose

Issues to consider when determining the management system "rigor" needed:

- Perception of the complexity, hazard, and risk involved with the process, the facility (or facilities), and the organization(s)
- Demand for the system results and the resources required to deliver them
- Current company/facility culture

Result: Design, correct, and improve PSM system activities

Table 8.2 Essential Features of a Good Safety Culture

1. Establish safety as a core value
2. Provide strong leadership
3. Establish and enforce high standards of performance
4. Formalize the safety culture emphasis/approach
5. Maintain a sense of vulnerability
6. Empower individuals to successfully fulfill their safety responsibilities
7. Defer to expertise
8. Ensure open and effective communications
9. Establish a questioning/learning environment
10. Foster mutual trust
11. Provide timely response to safety issues and concerns
12. Provide continuous monitoring of performance

8.2 AVOID PAST PSM SYSTEM FAILURE MODES

In most organizations, PSM system implementation and execution failures have occurred in the past. Similar failures can be anticipated and avoided by (1) evaluating PSM element failure modes and (2) determining PSM system failure causal factors.

8.2.1 Evaluate PSM Element Failure Modes

In order to evaluate past – and therefore possible future – PSM element failure modes, the following three-step process for evaluating each element is suggested:

1. Determine the basic PSM element steps by:
   a. Reviewing the element written programs
   b. Identifying element and system life-cycle activities completed and current status. In other words, which of the following stages is each element in?
      - Design and development
      - Implementation and rollout
      - Day-to-day utilization
      - Monitoring and improvement
   c. Developing (or reviewing and updating) the workflow diagrams of the element work process
   d. Reviewing relevant incident root causes that demonstrate element weaknesses or opportunities for improvement
   e. Reviewing relevant metrics for the element (leading and lagging indicators)
   f. Reviewing the results/findings from the previous two audit cycles for the element
   g. Assigning incidents, root causes, audit findings and observations, and metrics indicator performance to:
      - The life-cycle phase during which the element performance issue occurred
      - The workflow process point at which the element breakdown occurred
2. Highlight the element life-cycle phase where performance issues are greatest.
3. Highlight the workflow process point where most element performance issues have occurred.

The following example illustrates the concept of PSM element failure modes.

Example of MOC element failure modes

If we assume that most typical MOC programs have four life-cycle phases (design and development, implementation and rollout, utilization, and monitoring and improvement), then a knowledgeable evaluation team would likely identify the following potential failure modes (or similar ones) for the utilization phase:

- Failure to identify a proposed change, resulting in circumventing the system
- Change classified as an emergency change when it did not meet established criteria
- Mistakenly including a replacement-in-kind in the MOC review process

- Proposed change improperly classified, regarding either the type of MOC or the review path
- MOC origination information inadequate
- MOC initial review not completed or inadequate
- Inadequate MOC reviewers
- Wrong MOC review method used
- MOC hazard review path step missed, out of order, or incomplete
- MOC hazard evaluation inadequate, resulting in hazards missed or risks improperly evaluated
- Emergency MOC review procedure requirements not completed
- MOC authorization inadequate (e.g., wrong, missing, or risks accepted are inappropriate)
- Process safety information not updated based upon change
- Personnel not informed of change
- Personnel not trained on change
- Wrong or incomplete communication or training provided to personnel
- Temporary change left in place too long without further or periodic review
- Failure to restore system to original condition after a temporary change
- MOC review records inadequate or missing
- MOC delayed or lost in system

Similarly, failure modes can be developed for the other three MOC program life-cycle phases (i.e., design and development, implementation and rollout, and monitoring and improvement), keeping in mind that any phase could have the most gaps or errors.

8.2.2 Determine PSM Element or System Failure Causal Factors

PSM system or element failures can usually be tied to cultural causal factors. Therefore, failures can be either prevented or analyzed and corrected by:

- Evaluating PSM system or element failure causal factors by:
    - Linking (or "mapping) PSM system/element performance issues to culture features (see Appendix VII)
    - Comparing PSM system/element performance to known culture weaknesses
    - Identifying which culture features appear to be contributing to system/element performance lapses

- Ensuring sustainable PSM performance improvement by:
    - Making technical corrections to improve PSM element performance
    - Implementing culture improvement activities to address culture weaknesses
    - Monitoring culture changes and improvement

8.3 WATCH FOR EARLY WARNING SIGNS

The final step in effectively managing future process safety performance is to watch for early warning signs. If we use the analogy of driving a car into a ditch, where the ditch represents poor process safety performance, a company or site needs to be able to determine whether it is:

- in a process safety ditch,
- on the edge of the ditch,
- getting closer to or aiming at the ditch,
- moving away from the ditch, or
- maintaining proper distance from the ditch.

8.3.1 Be Alert for Organizational Warning Signs

In graphical terms, one way to recognize and react to organizational warning signs is to ensure that the organization's safety pyramid (see Chapter 2 and Figures 2.1 through 2.4 for background information) is not faulty, as described in Figure 8.1 (and in the related note below the figure explaining why it is faulty).

At a company level, these are some of the common warning signs that should be monitored:

- Organizational change/stress without sufficient PSM impact evaluation and mitigation, such as:
    - Externally induced changes related to regulations, enforcement, economics, disasters, being a merger and acquisition target, etc.
    - Internally induced changes due to loss of competency, loss of corporate memory, resources, loss of focus, initiative overload, mergers and acquisitions, leadership instability, demographics shifts, turnover, or absenteeism
- Loss of visibility and/or fidelity in performance evidence sources (i.e., not maintaining a good pyramid); this can result from poor reporting, trending, sharing, and/or monitoring of performance

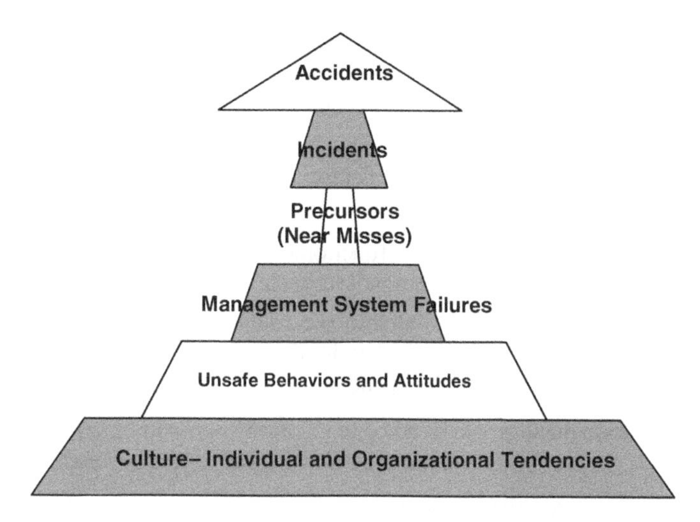

Figure 8.1 Example of a "Faulty" Safety Pyramid

> **NOTE**
>
> Why is this pyramid "faulty"? It is not well shaped or complete.
>
> Why? There has been a loss of visibility and/or fidelity in the performance evidence sources.
>
> What are examples of losses of visibility or fidelity?
>
> - Poor reporting (e.g., very few precursors, such as near misses, are being reported, and there should be more data [i.e., more events] at each lower level in the pyramid)
> - Poor trending (e.g., data are not being reported in a way that reveals the pyramid is misshapen)
> - Poor sharing (e.g., data are available, but are not being shared with the right people within the organization; i.e., the ones who can address the issue[s])
> - Poor monitoring (e.g., data at one level are not being collected, data are not being collected frequently enough, or the right people are not following the results)

- Signs of a poor process safety culture, such as:
  - People failing to report (or even hiding) issues related to poor performance or failures
  - A propensity to "kill the messengers" (i.e., those who report poor performance or failures are overtly or covertly punished, or openly criticized)

- Failure to question and learn (see the learning organization discussion in Section 8.1.1 above)
- Procedures not followed with no consequences evident
- Mixed and/or improper messages regarding the priority of safety vs. production
- Complacency (e.g., loss of sense of vulnerability)
- Low level of mutual trust
- "Silo" mentality where functional groups do not share information, work together on common issues, etc.
- Misplaced safety ownership (i.e., safety is a group or person's job, rather than the job of each individual)
- Invisible, ineffective leadership
- Normalization of deviance (e.g., alarms ignored, interlocks bypassed, a practice that is not consistent with procedures becomes accepted)

Similarly, these are the warning signs that should be monitored at the facility level:

- PSM leading indicators (see Chapter 5 of this book, *Guidelines for Risk Based Process Safety, Recognizing Catastrophic Incident Warning Signs in the Process Industries* [Ref. 8.10], and ANSI/API RP 754 [Ref. 8.11] for more information)
- Behaviors including:
  - Safe work practice nonconformance rates (e.g., the percent of nonconformance during behavioral observations, safe work practice field audits, etc.)
  - At-risk behavior or behavioral safety observation trend
  - Improper safety system bypass rate
  - Operating procedure or operating envelope deviations
- Company warning signs (as listed above), but applied at a local level:
  - Organizational
  - Visibility
  - Culture

8.4 CONSIDER OTHER ENHANCEMENTS

Finally, organizations should be alert to taking advantage of other enhancements to improve their PSM systems and reduce incidents and injuries. Some examples include:

- Increased and/or improved usage of information technology and software to improve the efficiency and/or effectiveness of the PSM systems (the "PSM Software Compilation" on the files on the Web accompanying this book lists many currently available software packages)
- The Serious Injury and Fatality (SIF) Prevention concept and similar approaches to address areas that may not be addressed by process safety programs (see Appendix II)
- Risk registers and similar approaches to manage and reduce site and/or enterprise risk
- Future publications from CCPS and other organizations that continue to advance the art and science of process safety

Beyond these examples, organizations should be continuously alert for new ideas from internal and external sources that could enhance their process safety programs.

8.5 REFERENCES

8.1 Center for Chemical Process Safety of the American Institute of Chemical Engineers, *Guidelines for Risk Based Process Safety*, John Wiley & Sons, Inc., Hoboken, New Jersey, 2007.

8.2 Center for Chemical Process Safety of the American Institute of Chemical Engineers, *Conduct of Operations and Operational Discipline for Improving Process Safety in Industry*, John Wiley & Sons, Inc., Hoboken, New Jersey, 2011.

8.3 Senge, Peter M., *The Fifth Discipline: The Art & Practice of the Learning Organization*, Doubleday, New York, New York, 2006.

8.4 Easterby-Smith, Mark, Luis Araujo, and John G. Burgoyne, *Organizational Learning and the Learning Organization: Developments in Theory and Practice*, SAGE Publications Ltd., Thousand Oaks, California, 1999

8.5 Garvin, David A., Amy C. Edmondson, and Francesca Gino, "Is Yours a Learning Organization?," *Harvard Business Review*, Harvard Business School Publishing Corporation, Boston, Massachusetts, March 2008.

8.6 Arendt, Steve, "Continuously Improving PSM Effectiveness – A Practical Roadmap," *Process Safety Progress*, American Institute of Chemical Engineers, New York, New York, Vol. 25, Issue 2, June 2006, pp. 86-93.

8.7 Throness, Barry, "Keeping the Memory Alive, Preventing Memory Loss That Contributes to Process Safety Events," *Process Safety Progress*, American Institute of Chemical Engineers, New York, New York, Vol. 33, Issue 2, June 2014, pp. 115-123.

8.8 Senge, Peter M., Art Kleiner, Charlotte Roberts, Richard Ross, George Roth, and Bryan Smith, *The Dance of Change: The Challenges to Sustaining Momentum in a Learning Organization*, Crown Business, The Crown Publishing Group, New York, New York, 1999.

8.9 Frank, W. L., "Safety Culture in the CCPS Risk-based Process Safety Model – An Update," Global Congress on Process Safety, 2014.

8.10 Center for Chemical Process Safety of the American Institute of Chemical Engineers, *Recognizing Catastrophic Incident Warning Signs in the Process Industries*, John Wiley & Sons, Inc., Hoboken, New Jersey, 2011.

8.11 ANSI/API Recommended Practice 754, *Process Safety Performance Indicators for the Refining and Petrochemical Industries*, American Petroleum Institute, Washington D.C., April 2010, www.publications.api.org/.

APPENDIX I: GLOBAL PSM REGULATIONS/GOOD INDUSTRY PRACTICES

Country (Region)	Title	Type REG or GIP	Summary	Link	Entry into Force Date	Implementation/ Compliance Date	Comparison to OSHA PSM	Comments
Global	*Risk Based Process Safety*	GIP	A 20-element process safety management system	http://www.aiche.org/ccps/resources/publications/books/guidelines-risk-based-process-safety	2007	—	Suggests a broader and deeper program than OSHA PSM and includes "soft" elements such as process safety culture	Center for Chemical Process Safety (CCPS) book
	High level framework for process safety management	GIP	Safety management framework capturing industry good practice To provide a consistent and effective approach for discussions with regulators, insurers, and other stakeholders	https://www.energyinst.org/technical/PSM/PSM-framework	Dec 2010		Comparable to CCPS	Developed by UK energy industry practitioners (incl. BP, Chevron, ConocoPhillips, E.ON, Marsh, Total, UK-HSE, UKPIA). Published by Energy Institute

Country (Region)	Title	Type REG or GIP	Summary	Link	Entry into Force Date	Implementation/ Compliance Date	Comparison to OSHA PSM	Comments
Americas								
Argentina								There are Labor Laws, but no process safety regulations
Brazil	NR-20 (Flammable and Combustible Liquids)	REG	This NR sets minimum requirements for the management of Health and Safety at Work against the factors of risks of accidents from activities of extraction, production, storage, transfer, handling, and manipulation of flammable and combustible liquids	http://portal.mte.gov.br/data/files/FF80808147596147014759F3612D634A/NR-20%20(atualizada%202014)%20(com%20prorroga%20prazos%20Prt.%201.079_14).pdf	June 1978	July 2014 (latest revision)	The latest NR-20 revision made this standard similar to OSHA PSM	There are no PSM regulatory agencies like the US. The regulatory body is the Brazilian Ministry of Labor that issues "Normas Regulamentadoras – NRs."
	NR-13 (Pressure Vessels and Piping)	REG	Pressurized vessels and piping code	http://portal.mte.gov.br/data/files/FF80808147596147014764A4E1D14497/NR-13%20(Atualizada%202014).pdf	June 1978	April 2014 (latest revision)	N/A	

Country (Region)	Title	Type REG or GIP	Summary	Link	Entry into Force Date	Implementation/ Compliance Date	Comparison to OSHA PSM	Comments
Americas								
Argentina								There are Labor Laws, but no process safety regulations
Brazil	NR-20 (Flammable and Combustible Liquids)	REG	This NR sets minimum requirements for the management of Health and Safety at Work against the factors of risks of accidents from activities of extraction, production, storage, transfer, handling, and manipulation of flammable and combustible liquids	http://portal.mte.gov.br/data/files/FF8080814759614701475 9F3612D634A/NR-20%20(atualizada%202014)%20(com%20prorroga%20prazos%20Prt.%201.079_14).pdf	June 1978	July 2014 (latest revision)	The latest NR-20 revision made this standard similar to OSHA PSM	There are no PSM regulatory agencies like the US. The regulatory body is the Brazilian Ministry of Labor that issues "Normas Regulamentadoras – NRs."
	NR-13 (Pressure Vessels and Piping)	REG	Pressurized vessels and piping code	http://portal.mte.gov.br/data/files/FF80808147596147014764A4E1D14497/NR-13%20(Atualizada%202014).pdf	June 1978	April 2014 (latest revision)	N/A	

Country (Region)	Title	Type REG or GIP	Summary	Link	Entry into Force Date	Implementation/ Compliance Date	Comparison to OSHA PSM	Comments
Americas								
Argentina								There are Labor Laws, but no process safety regulations
Brazil	NR-20 (Flammable and Combustible Liquids)	REG	This NR sets minimum requirements for the management of Health and Safety at Work against the factors of risks of accidents from activities of extraction, production, storage, transfer, handling, and manipulation of flammable and combustible liquids	http://portal.mte.gov.br/data/files/FF8080814759614701475 9F3612D634A/NR-20%20(atualizada%202014)%20com%20prorroga%20prazos%20Prt.%201.079_14).pdf	June 1978	July 2014 (latest revision)	The latest NR-20 revision made this standard similar to OSHA PSM	There are no PSM regulatory agencies like the US. The regulatory body is the Brazilian Ministry of Labor that issues "Normas Regulamentadoras – NRs."
	NR-13 (Pressure Vessels and Piping)	REG	Pressurized vessels and piping code	http://portal.mte.gov.br/data/files/FF80808147596147014764A4E1D14497/NR-13%20(Atualizada%202014).pdf	June 1978	April 2014 (latest revision)	N/A	

Country (Region)	Title	Type REG or GIP	Summary	Link	Entry into Force Date	Implementation/ Compliance Date	Comparison to OSHA PSM	Comments
Americas								
Argentina								There are Labor Laws, but no process safety regulations
Brazil	NR-20 (Flammable and Combustible Liquids)	REG	This NR sets minimum requirements for the management of Health and Safety at Work against the factors of risks of accidents from activities of extraction, production, storage, transfer, handling, and manipulation of flammable and combustible liquids	http://portal.mte.gov.br/data/files/FF8080814759614701475 9F3612D634A/NR-20%20atualizada%202014%20com%20prorroga%20prazos%20Pn.%201.079_14).pdf	June 1978	July 2014 (latest revision)	The latest NR-20 revision made this standard similar to OSHA PSM	There are no PSM regulatory agencies like the US. The regulatory body is the Brazilian Ministry of Labor that issues "Normas Regulamentadoras – NRs."
	NR-13 (Pressure Vessels and Piping)	REG	Pressurized vessels and piping code	http://portal.mte.gov.br/data/files/FF80808147596147014764A4E1D14497/NR-13%20(Atualizada%202014).pdf	June 1978	April 2014 (latest revision)	N/A	

Country (Region)	Title	Type REG or GIP	Summary	Link	Entry into Force Date	Implementation/ Compliance Date	Comparison to OSHA PSM	Comments
	Canadian Environmental Protection Act – Environmental Emergency Regulation, Section 200, Part 8	REG		http://www.ec.gc.ca/lcpe-cepa/default.asp?lang=En&n=26A03BFA-1	1999		N/A	
	Responsible Care	GIP	The Process Safety Network (PSN) is a CIAC committee. It is a volunteer (sweat equity) group developing tools and standards for process safety to ensure it maintains a high profile within member companies. The PSN also shares process safety incidents and knowledge gained	http://www.canadianchemistry.ca/ResponsibleCareHome.aspx		—	N/A	Chemical Industry Association of Canada (CIAC)

Country (Region)	Title	Type REG or GIP	Summary	Link	Entry into Force Date	Implementation/ Compliance Date	Comparison to OSHA PSM	Comments
	Canadian Society for Chemical Engineering (CSChE)	GIP	The PSM division of the CSChE is a national network of more than 300 volunteers. The division connects individuals with a common interest and expertise in PSM to create networking opportunities and to share ideas	CSChE *PSM Guide*, 4th edition				
Colombia								No process safety regulations at present
Mexico	NOM -028-STPS-2012	REG	Requires 16 PSM elements	http://trabajoseguro.stps.gob.mx/trabajoseguro/boletines%20anteriores/2012/bol044/vinculos/PROY-MOD-NOM-028.pdf	30 August 2012	28 February 2014	Very similar regulation to OSHA PSM	An unofficial translation of this regulation is provided in the files on the Web

Country (Region)	Title	Type REG or GIP	Summary	Link	Entry into Force Date	Implementation/ Compliance Date	Comparison to OSHA PSM	Comments
	Integral Security and Environmental Management System (SIASPA)	REG			1998		N/A	
	SSPA							
United States of America	29 CFR 1910.119 (OSHA)	REG	Requires 14 PSM elements for processes with listed toxics above listed threshold quantities (TQs) and/or with flammables >10,000 pounds	http://www.ecfr.gov/cgi-bin/text-idx?c=ecfr&SID=5df56ebb48b85542b53853863f60c639&rgn=div8&view=text&node=29:5.1.1.1.8.8.33.13&idno=29	24 February 1992	26 May 1992	—	
	40 CFR 68 (EPA)	REG	Requires most of the same OSHA PSM ("Prevention") elements + registration, offsite consequence assessment, and emergency response for listed toxics and flammables above listed TQs	http://www.ecfr.gov/cgi-bin/text-idx?c=ecfr&SID=5df56ebb48b85542b53853863f60c639&rgn=div5&view=text&node=40:16.0.1.1.5&idno=40	20 June 1996	21 June 1999	See Summary	

Country (Region)	Title	Type REG or GIP	Summary	Link	Entry into Force Date	Implementation/ Compliance Date	Comparison to OSHA PSM	Comments
	Responsible Care – Process Safety Code	GIP	Less specific, but most demanding in terms of expectations for continuous improvement	http://responsiblecare.americanchemistry.com/Responsible-Care-Program-Elements/Process-Safety-Code/Responsible-Care-Process-Safety-Code-PDF.pdf			N/A	American Chemistry Council
	API RP 75 *(Recommended Practice for Development of a Safety and Environmental Management Program for Offshore Operations and Facilities)*	GIP		http://www.techstreet.com/api/products/1157045	Reaffirmed May 2008		Includes elements similar to OSHA PSM	This recommended practice is currently under review by committee for update

Country (Region)	Title	Type REG or GIP	Summary	Link	Entry into Force Date	Implementation/ Compliance Date	Comparison to OSHA PSM	Comments
	Safety and Environmental Management Systems (SEMS) [Title 30 CFR Part 250 Subpart S]	REG		http://www.ecfr.gov/cgi-bin/text-idx?SID=73ecf0ca1763f93e3652e05e49894b7a&node=30:2.0.1.2.3.19&rgn=div6	October 2011		N/A	
	Chemical Facility Anti-Terrorism Standards (CFATS), Title 6, Chapter 1, Part 27	REG	This rule establishes risk based performance standards for the security of our nation's chemical facilities. It requires covered chemical facilities to prepare security vulnerability assessments, which identify facility security vulnerabilities, and to develop and implement site security plans, which include measures that satisfy the identified risk based performance standards	http://www.ecfr.gov/cgi-bin/text-idx?tpl=/ecfrbrowse/Title06/6cfr27_main_02.tpl			N/A	

Country (Region)	Title	Type REG or GIP	Summary	Link	Entry into Force Date	Implementation/ Compliance Date	Comparison to OSHA PSM	Comments
	Maritime Security: Facilities (MARSEC), Title 33, Chapter 1, Subchapter H, Part 105	REG	The Coast Guard employs a three-tiered system of Maritime Security (MARSEC) Levels. MARSEC Levels are set to reflect the prevailing threat environment to the marine elements of the national transportation system, including ports, vessels, facilities, and critical assets and infrastructure located on or adjacent to waters subject to the jurisdiction of the U.S. MARSEC Levels apply to vessels, Coast Guard-regulated facilities inside the U.S., and the Coast Guard	http://www.ecfr.gov/cgi-bin/text-idx?c=ecfr;sid=a485483870d82bafb395c519af3d4219;rgn=div5;view=text;node=33%3A1.0.1.8.53;idno=33;cc=ecfr	2003		N/A	

Country (Region)	Title	Type REG or GIP	Summary	Link	Entry into Force Date	Implementation/ Compliance Date	Comparison to OSHA PSM	Comments
Europe – EU								
Europe (EU)+UK	Seveso Directive 2012/18/EU (aka, "SEVESO III")	REG	Prevention of major accidents involving dangerous substances Categorization on the amount of dangerous substances present in lower and upper tier establishments Deploying a major accident prevention policy Information on seven-element safety management system to prevent major accidents Producing a safety report for upper-tier establishments Producing internal emergency plans for upper tier establishments Providing information in case of accidents	http://ec.europa.eu/environment/seveso/	4 July 2012	1 June 2015	Seven-element major accident prevention policy and safety management system: (i) organization and personnel, (ii) identification and evaluation of major hazards, (iii) operational control, (iv) management of change, (v) planning for emergencies, (vi) monitoring performance, and (vii) audit and review Safety report for upper tier establishments is a totally different approach	

Country (Region)	Title	Type REG or GIP	Summary	Link	Entry into Force Date	Implementation/ Compliance Date	Comparison to OSHA PSM	Comments
			Public concerned needs to be consulted and involved in the decision-making for investment projects					

Country (Region)	Title	Type REG or GIP	Summary	Link	Entry into Force Date	Implementation/ Compliance Date	Comparison to OSHA PSM	Comments
	EU Directive 2013/30/EU on safety of offshore oil and gas operations		Countries to appoint "Competent Authorities" to assess technical and financial capability of licensees, including ability to cover potential liabilities Operator must be approved by licensing authority Operators to submit a safety report detailing: • corporate major accident prevention policy, • safety and environmental management system, • report on major hazards, and • emergency response plans	http://ec.europa.eu/energy/oil/offshore/standards_en.htm and http://eur-lex.europa.eu/legal-content/EN/TXT/PDF/?uri=CELEX:32013L0030&from=EN	12 June 2013	18 July 2015	See above	Brings offshore closer to COMAH

Country (Region)	Title	Type REG or GIP	Summary	Link	Entry into Force Date	Implementation/ Compliance Date	Comparison to OSHA PSM	Comments
	ATEX 137 Workplace Directive (*Directive 1999/34/EC*) (aka, "ATEX 137")	REG	Minimum requirements for improving the safety and health protection of workers potentially at risk from explosive atmospheres. Responsibilities of employers and not manufacturers	http://eur-lex.europa.eu/LexUriServ/LexUriServ.do?uri=OJ:L:2000:023:0057:0064:EN:PDF	16 December 1999	30 June 2003	N/A	In parallel to the ATEX Directive (UK: DSEAR, GER: BetrSichV)
	Machinery Directive 2006/42/EC	REG	Harmonization of essential health and safety requirements for machinery at EU level. The definition machinery may also enclose assemblies of machinery up to industrial processes. Promotes the free movement of machinery within the Single Market; guarantees a high level of protection for EU workers and citizens	http://ec.europa.eu/growth/sectors/mechanical-engineering/machinery/index_en.htm	17 May 2006	29 June 2008	N/A	Replaced Machinery Directive 98/37/EC

Country (Region)	Title	Type REG or GIP	Summary	Link	Entry into Force Date	Implementation/ Compliance Date	Comparison to OSHA PSM	Comments
	PED (Pressure Equipment Directive, *PED Directive 97/23/EC*)	REG	Harmonization of essential health and safety requirements for equipment subject to a pressure hazard	http://ec.europa.eu/enterprise/sectors/pressure-and-gas/documents/ped/index_en.htm	29 May 1997	29 May 2002	N/A	The Directive is implemented in Europe by means of each national authority transposing its provisions into its legislation (e.g., PSSR in the UK)
	Registration, Evaluation, Authorisation & Restriction of Chemicals (REACH)	REG	Manufacturers or importers of >1 T/yr. of substances must register them with European Chemicals Agency. The registration package must be supported by an [extensive] standard set of data on that substance. No data, no market!	http://www.hse.gov.uk/reach/index.htm and http://echa.europa.eu/web/guest		1 June 2007	N/A	The Directive is implemented in Europe by means of each national authority transposing its provisions into its legislation (e.g., REACH in the UK)

Country (Region)	Title	Type REG or GIP	Summary	Link	Entry into Force Date	Implementation/ Compliance Date	Comparison to OSHA PSM	Comments
	CLP/GHS - Classification, labelling and packaging of substances and mixtures	REG	Aligns EU to the GHS (Globally Harmonised System of Classification and Labelling of Chemicals), a United Nations system to identify hazardous chemicals and to inform users about these hazards through standard symbols and phrases on the packaging labels and through safety data sheets (SDS)	http://ec.europa.eu/enterprise/sectors/chemicals/classification/index_en.htm		31 May 2015	N/A	The Directive is implemented in Europe by means of each national authority transposing its provisions into its legislation (e.g. COSHH and REACH in the UK)
All of the above EU Directives are separately implemented, in each of the 28 countries of the EU All national implementation texts are scrutinized, by the EU Commission, to ensure they will actually implement all the measures required in the directive								
Belgium		GIP	Guidance document for SEVESO	Appendix: http://www.lne.be/themas/veiligheidsrapportage/rlbvr/bestanden-rlbvr/tr/vr_rlbvr_rl_hbff_background_information_EN.pdf			See SEVESO above	And all EU regulations

Country (Region)	Title	Type REG or GIP	Summary	Link	Entry into Force Date	Implementation/ Compliance Date	Comparison to OSHA PSM	Comments
Netherlands	???	REG		Risk assessment manual: http://www.rivm.nl/dsresource?objectid=rivmp:22450&type=org&disposition=inline			See SEVESO above	And all EU regulations
Norway								
Sweden								
Switzerland								
UK	COMAH	REG	Implementation of EU's Seveso Directive	1999 COMAH: http://www.legislation.gov.uk/uksi/1999/743/contents/made 2005 Amendments: http://www.hse.gov.uk/comah/background/summary.pdf 2015 for Seveso III: http://www.hse.gov.uk/seveso/index.htm		1 June 2015	See SEVESO above	

Country (Region)	Title	Type REG or GIP	Summary	Link	Entry into Force Date	Implementation/ Compliance Date	Comparison to OSHA PSM	Comments
	DSEAR	REG	UK implementation of EU's ATEX regulations	http://www.hse.gov.uk/fireandexplosion/dsear.htm	2005		N/A	Defines area classification & equipment specifications Also covers Hierarchy of Control
	R2P2 (Reducing Risks and Protecting People)	GIP	Definitive basis for risk regulations	http://www.hse.gov.uk/risk/theory/r2p2.htm	2001		N/A	Basis of UK's risk approaches, defining "tolerability," ALARP, cost-benefit approach, etc.
	Environmental Risk Tolerability for COMAH Establishments	GIP	Quantitative definition of a Major Accident to the Environment; definition of tolerability thresholds for different receptors and risk based approach for ALARP demonstration for environmental risks	http://www.hse.gov.uk/aboutus/meetings/committees/cif/environmental-risk-assessment.pdf	2014	2015	First global quantitative approach for environmental risk assessments to bring environment to same level as people risks	Mandatory approach to be included in COMAH safety reports from 2015 for all UK sites

Country (Region)	Title	Type REG or GIP	Summary	Link	Entry into Force Date	Implementation/ Compliance Date	Comparison to OSHA PSM	Comments
	ALARP suite of guidance	GIP	Extensive HSE guidance on how to demonstrate ALARP has been achieved (including guidance on cost-benefit analyses)	http://www.hse.gov.uk/risk/expert.htm			N/A	Good training material
	Guidance on ALARP decisions in COMAH and HID's approach to ALARP decisions	GIP	Further guidance on ALARP demonstration	http://www.hse.gov.uk/foi/internalops/hid_circs/permissioning/spc_perm_37/ and http://www.hse.gov.uk/foi/internalops/hid_circs/permissioning/spc_perm_39.htm			N/A	
	COSHH, CLP, and REACH	REG	Implementation of EU regulations in the UK concerning Control of Substances hazardous to Human Health, Classification Labelling and Packaging and REACH	http://www.hse.gov.uk/coshh/detail/coshh-clp-reach.htm			N/A	Fist link for REACH

Country (Region)	Title	Type REG or GIP	Summary	Link	Entry into Force Date	Implementation/ Compliance Date	Comparison to OSHA PSM	Comments
	SI 2005 No. 3117, *The Offshore Installations (Safety Case) Regulations 2005*	REG	The main aim of the regulations is to reduce the risks from major accident hazards to the health and safety of those working on offshore installations or in connected activities. The regulations implement the central recommendation of Lord Cullen's report on the public inquiry into the Piper Alpha disaster: that the operator or owner of every offshore installation should be required to prepare a safety case and submit it to HSE for acceptance	http://www.legislation.gov.uk/uksi/2005/3117/pdfs/uksi_20053117_en.pdf	April 6, 2006	April 6, 2006	As indicated by the title, this is a different, safety case approach	

Country (Region)	Title	Type REG or GIP	Summary	Link	Entry into Force Date	Implemen-tation/ Compliance Date	Comparison to OSHA PSM	Comments
Europe – Non-EU								
Russia	PB- 03-517-02	REG					N/A	There is a general industrial safety regulation
Asia and Middle East								
Abu Dhabi	ADNOC-CoP-V5-06	REG		http://www.adnoc.ae/content.aspx?newid=136&mid=136				
Australia	NOHSC:1014 (National Standard for Control of Major Hazard Facilities)	REG		http://www.safeworkaustralia.gov.au/sites/SWA/about/Publications/Documents/271/NationalStandard_ControlMajorHazardFacilities_NOHSC_1014-2002_PDF.pdf		2002	Similar to the UK, this is a safety case approach	

Country (Region)	Title	Type REG or GIP	Summary	Link	Entry into Force Date	Implementation/ Compliance Date	Comparison to OSHA PSM	Comments
	WHS (Work Health and Safety)	REG	New WHS laws commenced on 1 January 2012 in many states and territories to harmonise occupational health and safety (OH&S) laws across Australia WHS legislation includes a model WHS Act, regulations, Codes of Practice, and a national compliance and enforcement policy	http://www.safeworkaustralia.gov.au/sites/SWA/about/Publications/Documents/598/Model_Work_Health_and_Safety_Bill_23_June_2011.pdf		2011	N/A	WHS is not really a PSM regulation. No direct reference to process systems or handling/storage of hazardous materials

Country (Region)	Title	Type REG or GIP	Summary	Link	Entry into Force Date	Implementation/ Compliance Date	Comparison to OSHA PSM	Comments
China	AQ/T3034-2010 (*Guidelines for Process Safety for Petrochemical Corporations*)	REG*	This China process safety guide or rule is issued by State Administration of Work Safety (SAWS) of China. SAWS is more like OSHA of US. Beside this rule, SAWS now recommends that HAZOP be used for newly built facilities, especially in government-owned company. It seems someday in near future SAWS will issue a dedicated rule for HAZOP		May 1, 2011		Most items are copied from OSHA PSM, with very few items added	*It is our understanding that although this is titled as a "guideline," it would be enforced as a regulation if a foreign company has an event
India	Oil India Safety Regulations						N/A	Similar to UK HSE regulation
	(Production and Natural Gas – Offshore Regulation)						N/A	

Country (Region)	Title	Type REG or GIP	Summary	Link	Entry into Force Date	Implementation/ Compliance Date	Comparison to OSHA PSM	Comments
China	AQ/T3034-2010 (*Guidelines for Process Safety for Petrochemical Corporations*)	REG*	This China process safety guide or rule is issued by State Administration of Work Safety (SAWS) of China. SAWS is more like OSHA of US. Beside this rule, SAWS now recommends that HAZOP be used for newly built facilities, especially in government-owned company. It seems someday in near future SAWS will issue a dedicated rule for HAZOP		May 1, 2011		Most items are copied from OSHA PSM, with very few items added	*It is our understanding that although this is titled as a "guideline," it would be enforced as a regulation if a foreign company has an event
India	Oil India Safety Regulations						N/A	Similar to UK HSE regulation
	(Production and Natural Gas – Offshore Regulation)						N/A	

Country (Region)	Title	Type REG or GIP	Summary	Link	Entry into Force Date	Implementation/ Compliance Date	Comparison to OSHA PSM	Comments
Japan	High Pressure Gas Safety Act		The High Pressure Gas Safety Act, the Fire Services Act, the Industrial Safety and Health Law, and the Act on the Prevention of Disasters in Petroleum Industrial Complexes and Other Petroleum Facilities are collectively referred to as the "four safety acts."	e-gov link: http://law.e-gov.go.jp/cgi-bin/idxsearch.cgi High Pressure Gas Safety Act: http://www.cas.go.jp/jp/seisaku/hourei/data/HPGSA.pdf&rct=j&frm=1&q=&esrc=s&sa=U&ei=a0IXVK7hDay18gGGyoCICA&ved=0CBoQFjAB&usg=AFQjCNGwA9pCxOyR-lfhR1efRBr-c2ogg	2006		N/A	
Kuwait							N/A	No overarching regulation, but a number of "safety study" requirements
Malaysia	Occupational Safety and Health Act	REG		http://www.dosh.gov.my/index.php?lang=en	2006		N/A	

Country (Region)	Title	Type REG or GIP	Summary	Link	Entry into Force Date	Implementation/ Compliance Date	Comparison to OSHA PSM	Comments
New Zealand	Health and Safety in Employment Petroleum Exploration and Extraction Regulations 2013	REG			2013		N/A	Similar to Australia, a safety case approach is used
Oman		REG					N/A	Petrochemical Oman follows requirements similar to major oil companies
Saudi Arabia							N/A	Government-owned companies have extensive PSM procedures

Country (Region)	Title	Type REG or GIP	Summary	Link	Entry into Force Date	Implementation/ Compliance Date	Comparison to OSHA PSM	Comments
Singapore	Singapore Standard SS506 Part 3: 2013, Occupational Safety and Health (OSH) Management System – Requirements for the Chemical Industry	REG		**PREVIEW ONLY** http://www.singaporestandardseshop.sg/data/EcopyFileStore/061114131101Preview%20-%20SS%20506-3-2006.pdf http://rpsonline.com.sg/proceedings/9789810714451/html/GoHengHuat.pdf		2013	Some management system requirements, along with a safety case	The Ministry of Manpower expects a safety and health management system approach: http://www.mom.gov.sg/workplace-safety-health/safety-health-management-systems/Pages/default.aspx
	National Environment Agency	REG	One-time QRA report for new chemical plants	http://app2.nea.gov.sg/			N/A	
South Korea	Occupational Safety and Health Act	REG	In particular, Article 49-2 (Submission etc. of Process Safety Report) lays out requirements for submission, review/approval, and updating of such reports	http://english.kosha.or.kr/english/main.do	January 2012		N/A	

Country (Region)	Title	Type REG or GIP	Summary	Link	Entry into Force Date	Implementation/ Compliance Date	Comparison to OSHA PSM	Comments
Singapore	Singapore Standard SS506 Part 3: 2013, Occupational Safety and Health (OSH) Management System – Requirements for the Chemical Industry	REG		PREVIEW ONLY http://www.singaporestandardseshop.sg/data/EcopyFileStore/061114131101Preview%20-%20SS%20506-3-2006.pdf http://rpsonline.com.sg/proceedings/9789810714451/html/GoHengHuat.pdf		2013	Some management system requirements, along with a safety case	The Ministry of Manpower expects a safety and health management system approach: http://www.mom.gov.sg/workplace-safety-health/safety-health-management-systems/Pages/default.aspx
	National Environment Agency	REG	One-time QRA report for new chemical plants	http://app2.nea.gov.sg/			N/A	
South Korea	Occupational Safety and Health Act	REG	In particular, Article 49-2 (Submission etc. of Process Safety Report) lays out requirements for submission, review/approval, and updating of such reports	http://english.kosha.or.kr/english/main.do	January 2012		N/A	

Country (Region)	Title	Type REG or GIP	Summary	Link	Entry into Force Date	Implementation/ Compliance Date	Comparison to OSHA PSM	Comments
	Framework Plan on Hazardous Chemical Management	REG		http://eng.me.go.kr/eng/web/main.do	2005		N/A	
United Arab Emirates	Federal Law No. 8 (Regulations of Labour Relations)	REG		http://www.dsg.gov.ae/sitecollectionimages/content/pubdocs/uae_labour_law_eng.pdf&rct=j&frm=1&q=&esrc=s&sa=U&ei=dz8XVN7pAsao8QHC-oGoDQ&ved=0CCgQFjAD&usg=AFQjCNHCMOUJ0Y64MIbQU_h-dz8K3f0NNQ	1980		N/A	

APPENDIX II: ELI LILLY AND COMPANY PSM IMPLEMENTATION CASE STUDY

Preface and objectives of case study

This case study provides tools and implementation details, as it follows the approach taken by Eli Lilly and Company (Lilly), a global pharmaceutical firm, during implementation of its Process Safety Management (PSM) program.

Everything was beautiful and nothing hurt. Until he died.

Kurt Vonnegut – *Slaughterhouse-Five*

Kurt Vonnegut's quote seems appropriate when implementing PSM, particularly for a company like Lilly. Lilly was founded in and maintains its corporate headquarters in Indianapolis, Indiana – the same place that served as home to one of America's satirical, yet poignant writers. PSM is a high-consequence, low-frequency safety endeavor. As such, it can get overlooked until it's too late – sometimes tragically.

Implementing PSM is a journey; but in the end, catastrophic prevention is mostly about people. It means:

- getting people home alive and in good condition at the end of the workday,
- ensuring our neighbors stay alive and unharmed, and
- having a safe workplace where employees can return.

It doesn't get fundamentally more important than that.

At Lilly, the lessons learned were crucial to success in implementing and sustaining PSM. We determined we needed to do the following:

- Establish a PSM champion, with ongoing governance and oversight, at the senior management level
- Utilize credible, experienced PSM staff and HSE leadership with site experience
- Provide a simple reason to implement PSM to create a sense of vulnerability and a sense of urgency among people
- Recognize the need for various risk levels and ensure higher risk processes have additional safeguards/requirements (e.g., Safety Critical Operations – SCO concept)
- Measure progress during implementation

- Create practical tools to help the business (not just add-ons) and integrate them into existing business processes (e.g., HSE incident and change management)
- Establish a credible PSM audit program to help with sustainment
- Assist management with sustainment through fundamental leading and lagging metrics that sites can self-report
- Seek outside advice from credible consultants to help with communication, sustainment, and improvement initiatives among personnel and management
- Periodically reinvigorate PSM to keep it continuously alive

We hope the lessons learned at Lilly are helpful in implementing and sustaining PSM at other companies and sites.

The catastrophic prevention journey continues to this day at Lilly. The Lilly PSM program, Globally Integrated Process Safety Management (or GIPSM for short – phonetically pronounced jip'sem), was initially intended to cover approximately 80 percent of our catastrophic potential. As stated by our PSM leader at the time of implementation, "Don't let perfect get in the way of good."

To address the remaining 20 percent, Lilly recently launched two initiatives: the Catastrophic Potential Hazard (CPH) aspect initiative and the Serious Injury and Fatality (SIF) initiative. Tools and details for these initiatives are included in this case study.

The situation at Lilly (circa 1998)

In the late 1980s, Lilly's process safety-related activities (often informal) were scattered throughout manufacturing. At that time, in response to accidents such as the Bhopal disaster, manufacturing and development engineering management recognized the need for a more formal process safety management system. This resulted in a Process Hazard Review, now commonly known as a Process Hazard Analysis. However, shortly thereafter, federal regulations took effect and like most firms, Lilly implemented PSM along with regulatory United States PSM federal government requirements – OSHA Process Safety Regulation 29 CFR 1910.119. The approach was typical of most firms:

- A corporate program manual was written in the mid-90s by personnel with experience at other companies or experience at Lilly plant sites, but with limited experience working at each site that handled hazardous chemicals.
- Site feedback was often obtained from safety professionals and not from engineering and manufacturing management during PSM manual development.
- Assistance was provided in implementation through site visits, teleconferences, and email from corporate personnel.

- Audits were led by corporate personnel but did not include external expertise/learning.

Prior to 1995, plant sites conducted their own self-assessments and these assessments found limited gaps. In 1995 and 1996, Lilly conducted its first PSM audits. The audits found some major gaps that were not consistent with the previous self-assessments.

These audit findings led to the formation of the Process Safety Task Force consisting of site manufacturing management along with representatives of various corporate groups (HSE, Engineering, Development, etc.). The task force began meeting in 1995. The task force raised the visibility of PSM among management and moved some PSM efforts forward. But many people still viewed PSM as something they were being told to do (e.g., for compliance) rather than something they saw value in doing. One key recommendation of the task force was that a round of PSM audits should be conducted in late 1997 to check site progress in addressing the issues that had been raised in 1995 and 1996.

Prior to the 1997 audits, two events significantly changed the course of PSM at Lilly. A fire occurred in a chemical process research and development pilot plant. There were no injuries and the fire caused limited damage. However, operations were disrupted. Three days later a release of a hazardous chemical at another plant resulted in the plant being shut down for six months. Fortunately, neither site personnel nor its neighbors were exposed to the hazardous material from the release. These events prompted a refocusing of corporate planned audits.

The refocused corporate planned audits indicated fundamental flaws in the PSM program:

- At most sites, PSM was not understood by management or site personnel; at best, it was considered just another HSE program.
- Many chemicals not OSHA PSM covered, but with similar consequences, were not covered by the PSM program.
- PSM elements stood alone and were not integrated into other business processes at plant sites.
- Corporate PSM personnel were not seen as credible experts at plant sites.

Why PSM implementation was lacking in implementation up to that point

The flaws indicated by the PSM audits were fundamental reasons why the PSM implementation was lacking up to that point. These included a lack of:

- understanding and support from management;
- integration with business systems, creating extra work for personnel that often did not get executed;
- ineffective training and tools that created difficulty in understanding and executing requirements; and
- capable personnel to implement PSM.

The process where senior management identified the need and drove the change

The two events in 1997 at Lilly plant sites were a wake-up call for the PSM Task Force and other plant management. The plant site leader in charge of the PSM Task Force used his influence to take the learning from these events to other manufacturing management and restructured the efforts of the PSM Task Team.

This PSM Task Team was augmented by consultants both from within and outside the firm. They assisted manufacturing management in establishing the fundamentals for GIPSM at Lilly. These fundamentals are summarized in a series of GIPSM icons that were simple to understand by management and Lilly personnel. Shown and described below are the four GIPSM icons used during implementation:

- The Reason for GIPSM
- GIPSM Process Map
- Defense/Accident Model
- Production/Protection Space

The Reason for GIPSM

PSM events happen so infrequently; therefore, one of the greatest challenges in implementation is creating a sense of vulnerability. This icon gave a clear, easy-to-understand set of goals for GIPSM. It stresses what could happen in process safety incidents, thereby creating a sense of vulnerability.

This icon uses a training exercise photo of burning process equipment to create that sense of urgency. At the same time, this icon covers the need to comply with both internal and external regulations and requirements.

GIPSM Process Map

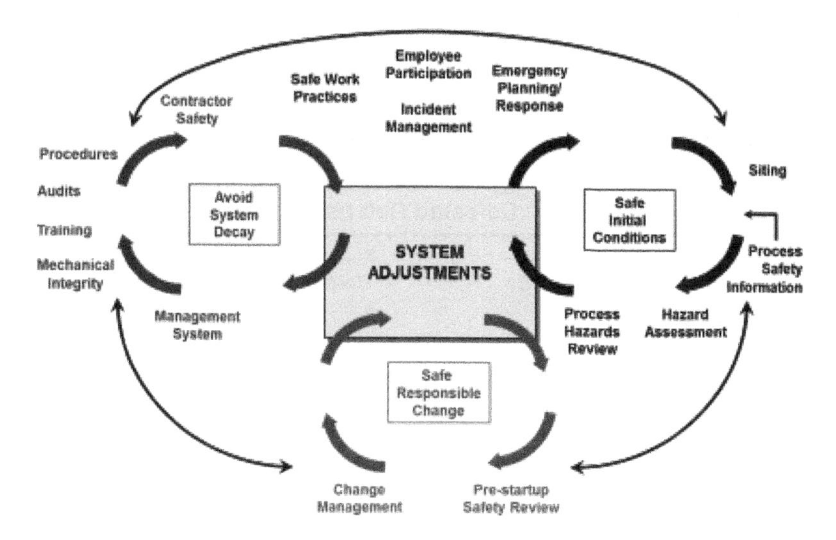

The GIPSM Process Map was jointly developed by the PSM Task Force (headed by a senior plant manager) and an internal senior consultant. It was finalized by senior leadership members of the PSM Task Team. This "Three Loop Mental Model" utilizes three systems to help ensure adequate defenses are built in, remain in place, and do not decay. Overlaid on this "Three Loop Mental Model" were the 16 GIPSM elements:

- Safe Initial Conditions – Facility Siting, Hazard Assessment, Process Safety Information, and Process Hazard Review
- Safe Responsible Change – Change Management and Pre-Startup Safety Review
- Avoid System Decay – Management System, Mechanical Integrity, Procedures, Training, Contractor Safety, and Audits
- Overarching Elements that Interact with All Loops/Elements – Incident Management, Employee Participation, Emergency Planning/Response, and Safe Work Practices (i.e., Hot Work, Confined Space Entry, Line-breaking, Lockout/Tagout, etc.)

The heart of this concept is that each of these elements follows the Plan-Do-Check-Adjust (PDCA) management concept, which interacts with your business systems/operations (i.e., Managing System Adjustments).

Note: This model is very flexible and can be used for other business processes (e.g., quality, engineering, operations) where Initial Conditions, Change, and System Decay are anticipated.

Defense/Accident Model

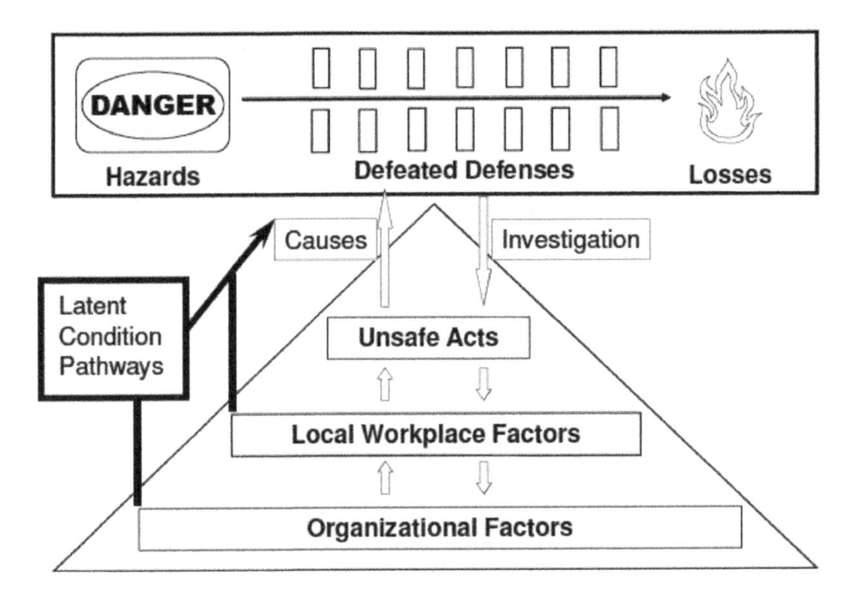

*** Adapted from *Managing the Risks of Organizational Accidents*, James Reason**

The Defense/Accident Model was first introduced by an internal senior technical resource. It basically states that defenses can be overcome to create losses. Typical defenses for a plant site include:

- Safe Design and Adequate Information,
- Operating Procedures and Training, and
- Maintenance and Change Management Programs.

The basic reasons why defenses become defeated are the final four components of the defense model:

- Unsafe Acts
- Local Workplace Factors

- Organizational Factors
- Hidden failure conditions (Latent Condition Pathways)

GIPSM's elements provide 16 management systems that bolster defenses by doing the following:

- Initially providing adequate defenses
    - Process Safety Information
    - Process Hazard Review
    - Hazard Assessment
    - Siting
- Ensuring defenses remain in place
    - Change Management
    - Pre-startup Safety Review
    - Management System
    - Audits
    - Employee Participation
- Ensuring defenses are operational
    - Mechanical Integrity
    - Contractor Safety
    - Training
    - Procedures
    - Safe Work Practices
- Ensuring consequences are mitigated and preventative learning takes place
    - Emergency Planning and Response
    - Incident Management

Production/Prevention Space

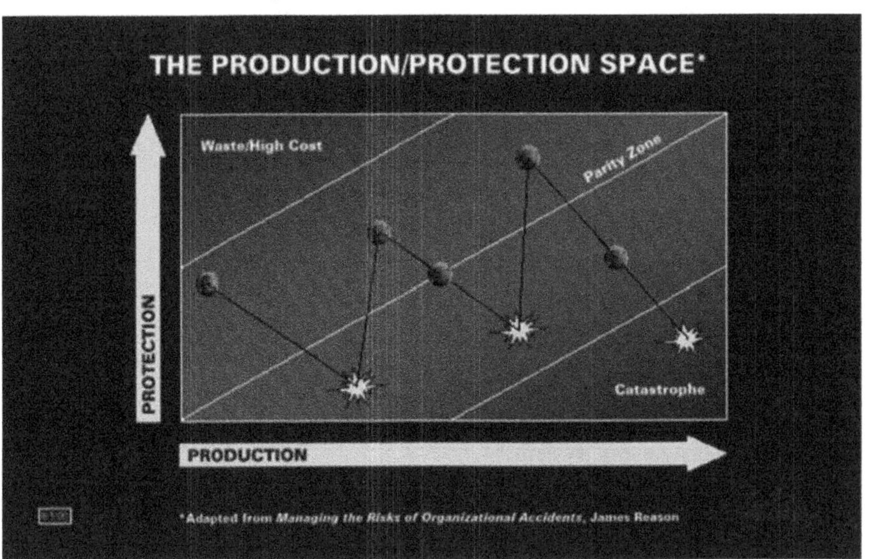

The Production/Protection Space Icon illustrates that businesses need to find a balance when considering additional safeguards for production processes. Plant sites need to find the "Parity Zone" where costs and the level of defenses for their processes are balanced, or catastrophic losses can occur.

This icon further shows that as risks increase in production (i.e., PSM processes), so does the need for additional safeguards. This concept provides a basis for considering various risk levels in PSM, as well.

Senior management supported the concept of creating risk categories for PSM chemical processes that had higher potential catastrophic consequences. The implementation team was asked to create a process to identify processes with higher risk. They created a Safety Critical Operations (SCO) Selection Tool (available in the files on the Web accompanying this book). This tool allowed sites to determine which PSM processes had a greater level of risk that could be considered Higher Risk Operations (HROs) and SCOs. The team conducted research and worked with sites to identify approximately 32 additional requirements for SCOs. Sixteen requirements were also eventually created from the SCO requirements and assigned to HRO processes. These additional requirements added additional safeguards (both management system and physical safeguards) for processes with higher risk associated with them.

All of the participation that laid the groundwork

Senior manufacturing management agreed to fund GIPSM implementation similar to a capital project, with defined goals and a project manager who worked at and understood Lilly manufacturing plant sites. He hired a staff of personnel with experience in PSM implementation and a consulting firm known for PSM implementation.

A conference was held to kick off the GIPSM initiative, with senior plant management as the key contributors to identifying the goals and laying the groundwork, including the following:

- Recognizing different risk levels and requirements to match
- Integrating PSM elements into existing business processes
- Creating tools that work with existing business process or business process needs
- Creating corporate requirements that would include all regulatory requirements at global sites
- Ensuring chemicals with similar consequences but not covered by regulations are covered by the same requirements
- Creating a credible gap analysis that incorporates plant sites' input
- Creating a time frame for implementing GIPSM at Lilly plant sites, with annual or biannual conferences and renewal events, as illustrated by the graphic on the next page

The risk based approach (applicability, low risk, Catastrophic Potential Hazard [CPH], high risk, HRO, SCO, etc.)

Senior management supported the concept that similar chemicals with catastrophic consequences – but not covered by PSM regulations – should follow GIPSM requirements. Consequently, they supported the creation of an Applicability GIPSM Process Risk Screening Tool. This tool has had numerous editions over the years, but its fundamental basis was the work completed by a senior Lilly engineer. This work included researching the basis for chemicals covered by the OSHA PSM standard, the EPA Risk Management Program (RMP) rule, and European "Seveso" directives. This applicability approach used that basis to create categories of chemicals with similar catastrophic consequences for Flammability, Explosion, Runaway Chemical Reaction, and Health Hazards. The files on the Web accompanying this book provides a Process Risk Screening Tool for your use as you feel appropriate.

Note: Instead of the Lilly Laboratory Safety Labeling Code, the NFPA Hazard Rating Code can be used in this tool.

Note: Dust explosion hazards are not included in the Process Risk Screening Tool, as Lilly has separate standards on combustible dusts.

GIPSM Next Level

Level 5	• PSM finally a routine part of the way we think, not separate. PSM concerns are incorporated from Research & Development process to facility design, to ongoing operations (i.e., Research & Development, Corporate Project Engineering, Process Engineering, Quality, etc.)	**2006+**
Level 4	• Focus on Execution (Safety Critical Operations Audits as an example) • Focus on Integration – everyone at the table (i.e., Research & Development scientists, corporate project engineers, process engineers, tech services scientists, Quality personnel, etc.) • Functional Leadership Support – not just Health and Safety or manufacturing management • Metrics focus on Execution – external auditors monitor progress	**2004 – 2006**
Level 3	• "Program" improvements in "Focus" areas • Metrics in place to monitor "program" improvement • Site Element Assessments/Near Miss/Incident Management/ Mechanical Integrity improvements measurable	**2001 – 2003**
Level 2	• "Program" in Place – basis for improvement • Safety Critical Operations requirements created • Some metrics identified for monitoring improvement	**2000 – 2001**
Level 1	• GIPSM created – requirements created by sites – gaps identified • Metrics for implementation of Gap Analysis at sites created – 1,000 of hours of work – program foundation	**1998 – 2000**
Level 0	• What PSM Program? – We don't have PSM incidents at Lilly • PSM Task Force created – composed of senior plant site and corporate management	**1993 – 1998**

The GIPSM approach created requirements for High Risk, HRO, and SCO processes that were intended to provide safeguards for approximately 80 percent of Lilly's process risks.

It was eventually recognized that there are other hazards, both chemical and physical, that have catastrophic potential, but without any specific visibility of this catastrophic risk or requirements for additional safeguards. These included processes which were just outside the threshold quantity for PSM, highly corrosive materials (acids and bases), high pressure steam processes, etc. Eventually senior management agreed to sponsor a team, including the Engineering Tech Center, to create the CPH Risk Screening Tool (also available in the files on the Web accompanying this book). This tool identifies both chemical and physical catastrophic processes and assists in declaring them as PSM aspects. Once identified, plant sites are expected to evaluate safeguards for these CPH processes and, if needed, create additional safeguards.

With the addition of the CPH aspect initiative, the graphic on the next page (the Process Risk Pyramid) represents the current risk based approach.

Examples of how GIPSM caused a step change and now has a continuous improvement aspect

The GIPSM implementation approach consisted of numerous phases:

- Creation of "As Is" and "Should Be" GIPSM Process Description, GIPSM Process Requirements, and GIPSM Maps for each GIPSM element by internal experts and external consultants. These were vetted by plant site personnel at a workshop to update and ensure inclusion of global requirements. The files on the Web accompanying this book provide an example of Pre-Startup Safety Review (PSSR) documents (PSSR Process Description document, along with the PSSR Process Map and PSSR Requirements Document).
- Creation of GIPSM tools for plant sites to use
    - Applicability – Process Risk Screening Tool, SCO Selection Tool
    - Process Safety Information – PSI Template Example*
    - Change Management – Change Management Response and Impact Assessment Tool Comments* used with automated Change Management database and tracking tool
    - Mechanical Integrity – Mechanical Integrity List* and Mechanical Integrity Test, and Inspection Guide*
    - Incident Management – Root Cause Analysis process used with automated Incident Management database and tracking tool (currently utilizing TrackWise)

(* Items in the files on the Web accompanying this book)

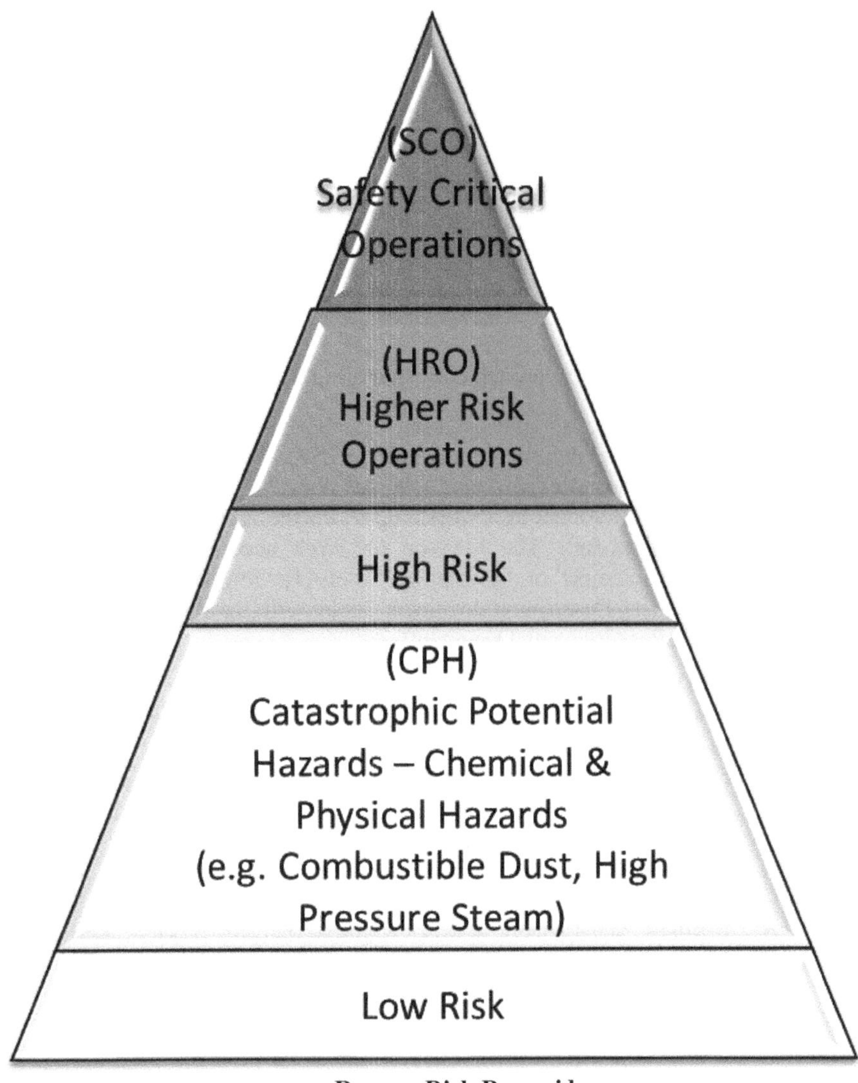

Process Risk Pyramid

- Creation of training courses for GIPSM
    - A two-day leader-led GIPSM Concepts Course (Exploring GIPSM Concepts, using case studies of external and internal incidents and human factors) intended for engineers, engineering, and operations management, and safety professionals

  - A half-day leader-led course introducing concepts and case studies and tailored for the plant site PSM program (i.e., plant site GIPSM chemicals/processes) intended for all maintenance, operations, and support personnel up to and including plant site supervisors and managers
  - A computer-based training course to introduce all PSM personnel to GIPSM and some high-level requirements of the program, and some content tailored for the plant site PSM program (i.e., plant site GIPSM chemicals/processes)
- Development of a "Brown Paper" As-Is/Should-Be Gap Analysis process. This included (1) placing large rolls of brown paper/butcher paper on walls and placing the Should-Be map on the paper, and then (2) working with a site team to map the business process steps at the plant site. When gaps were identified, an action plan was developed to address the gap.
- Sending implementation teams of at least three people to each plant site to conduct As-Is/Should-Be Gap Analyses at sites. Each site was required to supply at least three teams for up to four weeks at a time, with the purpose of creating action plans with sites that would integrate GIPSM with local site processes, owned by local personnel.
- Assistance to sites with implementation by helping them:
  - use tools,
  - use training templates,
  - organize site promotion events, and
  - create site communication plans (e.g., providing promotional and communication tools).
- Creation of GIPSM element process owners at sites (site element owners), who were charged with ensuring implementation of action plans. In addition, global element owners were created, intended to keep in touch with and provide continuous improvement expertise (including external benchmarking) for the site element owners.
- Creation of a communication plan with a long-term approach, which included conducting conferences annually or biannually. This provided continuous improvement support and reinvigorated GIPSM.
- Creation and ongoing support of a GIPSM Board of Directors, made up of senior site and manufacturing management, which proved key to sustainment. This board provided direction, resources, and continuous improvement support to the GIPSM program, while also reviewing key metrics.

Critical failures when implementing PSM

- Low turnover rate of PSM personnel and PSM leaders at both the corporate level and site level during implementation is critical. Most PSM personnel and PSM leaders were in their roles for three to four years or longer. If personnel know they will be in a PSM job for just a year or two, they may not be motivated to take responsibility for the success and long-term sustainment of PSM.
- Failure to recognize the people implementing PSM. PSM implementation is a very long process and a hard job, where success is measured by not having an event. You usually do not get credit for something not happening. That's why it is important to give awards or recognition as implementation progresses. To keep that drive high, some reinforcement or recognition is necessary. In addition, ensure that PSM is implemented at a site level by those who actually run things at a plant site. For example, we were having problems implementing at a site until the production and engineering team leaders were actually made leaders of GIPSM teams, responsible for implementing some elements.
- Implementing PSM is hard work. People come up with all kinds of excuses why they can't do something. We created a communication plan where we periodically created PSM Global Events to promote implementation. For example, we held PSM global conferences periodically so that we could network and have people see that other sites had the same problems and that they overcame those problems. We also created news articles on implementation status at each site and in our global communication systems. This helped share solutions and fostered networking between personnel at plant sites.

Items crucial to sustaining the program and continuous improvement

- A good, credible PSM audit program is key for both plant site personnel and management. You have to go out and verify that what the sites say they are doing, they are in fact doing. Consequently, the PSM audit program was redeveloped to include:
  - a comprehensive PSM audit protocol that included regulatory, SCO, and HRO requirements;
  - an outside consultant who helps maintain objectivity and ensures a nonbiased approach;
  - a representative from another plant site on the audit team to add plant site credibility;
  - a three-year frequency for PSM audits;
  - an additional SCO audit protocol and separate SCO audit for plant sites with SCO processes; and

  - o daily updates during the audit and for corporate management in order to resolve issues prior to the audit team completing the onsite audit (crucial to credibility).

- You need to have a few fundamental leading and lagging metrics that all sites self-report. These metrics must be made public *(between sites and management)*. The primary lagging metric was PSM events. The leading indicators were Mechanical Integrity Preventive Maintenance Plans Overdue, and both PHRs Overdue and PHR Action Plans Overdue.

  These metrics should be shared periodically with upper management in a public forum or team setting that has authority to act (e.g., a PSM Board of Directors meeting, composed of plant managers). It is also important to mandate that the organizational head require his site leaders to attend this sort of meeting, so that all the sites will think that PSM is really important.

- You need to have credible outside consultants and others come in and tell the senior leaders and site heads what it's like to have a major PSM event. Lilly had a senior leader from a major chemical company come and speak at a dinner. It was very sobering to hear him talk about the state police, EPA, local police, health department, coroner, etc. – all coming when they had a major PSM event. Then several years later, Lilly had the CEO of the company that had the dust explosion visit, and we did a live, interactive Business TV round table that was broadcast globally to our PSM sites. This visit really brought the PSM message home to site heads and senior leaders. *They did not want a PSM event to happen on their watch.*

Lilly hired a communication expert to help in that area. He was tasked with designing a communication plan that could be tailored to each plant site. Among the communication lessons learned were the following:

- Brand and package a message and communicate it over and over again. You can't overcommunicate the PSM messages (icons) – people only absorb 1 in 10 messages. Consider video messages, posters, email messages, etc.
- You need to create motivational and continuous improvement PSM implementation events/opportunities periodically (i.e., PSM brown paper fairs, communication fairs with personnel, conferences, video conferences, periodic speakers, etc.). Promotional items displaying the PSM brand as giveaways can help (i.e., pens, coffee mugs, t-shirts, etc.).
- You need to create a network of PSM personnel who are willing to become experts in PSM elements and share their experiences in teleconferences, video conferences, etc.
- You need to create consistent training templates and courses that motivate and engage people to get that feeling of vulnerability and gain the urgency to continuously implement PSM.

Future of the Lilly Catastrophic Prevention Journey and Summary

As stated previously, the creation of the Catastrophic Potential Hazard (CPH) aspect requirement and the Serious Injury and Fatality (SIF) initiative are an attempt to address the 20 percent of catastrophic risk that GIPSM did not address.

The Lilly CPH aspect requirement consists of the following phases:

- Conducting an analysis of site process hazards using tools such as the CPH Risk Screening Tool (available in the files on the Web accompanying this book) created by the Lilly technical experts
- Identifying and declaring the CPH processes from the tool as aspects at sites, with strengthened safeguards to ensure prevention of catastrophic incidents

A plant site at Lilly has created a CPH program that incorporates some GIPSM requirements:

- Conducting a CPH Assessment (e.g. Process Hazard Review, FMEA, LOPA, etc.) for identified CPH processes
- Requiring Process Safety Information for the CPH Assessment
- Requiring Procedures and Training for identified CPH processes, driven by a CPH Assessment
- Requiring Change Management for identified CPH processes
- Requiring Preventative Maintenance for CPH process equipment
- Requiring Root Cause Analysis for CPH process incidents

The Lilly SIF initiative consists of the following:

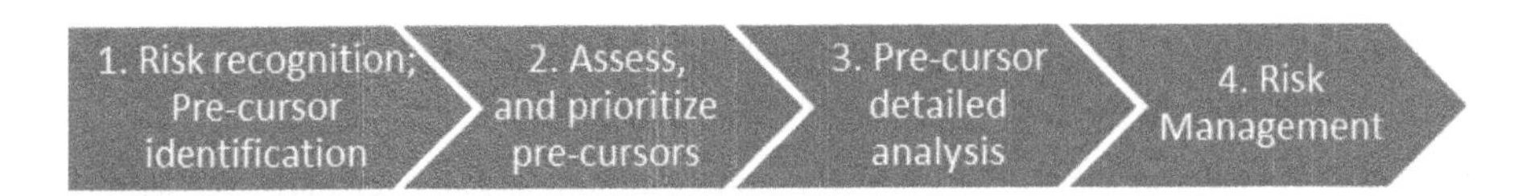

The heart of this initiative is to identify SIF Pre-Cursors. A SIF Pre-Cursor is defined as "a high risk situation, in which management controls are either absent, ineffective, or are not complied with and if allowed to continue or repeat could reasonably result in a serious injury or fatality."

An analysis of Lilly injuries and near misses at manufacturing operations has yielded a list of 41 SIF Pre-Cursors grouped into eight SIF Pre-Cursor categories:

- Working from Heights
- Whole Body Contact with Energized Equipment
- Confined Space Entry

- Heavy Equipment and Forklift/Powered Industrial Truck (PIT) Operations
- Electrically Classified Area/High Voltage Electrical Work
- Fires and Explosions
- Falling Object Potential
- Bulk Toxic Chemical Handling

These Pre-Cursors are listed in the Serious Injury and Fatality (SIF) Categories and Pre-Cursors Reference Guide (available in the files on the Web accompanying this book).

The objective of this initiative is to conduct global detailed analyses, with the goal of determining the critical requirements that prevent these Pre-Cursors from occurring. The intent is then to share these critical requirements with plant sites and work with them to ensure that gaps with these critical requirements are resolved at plant sites.

Note: The SIF Initiative marries Process Risk with Occupational Safety Behavioral Risk. As such, it has the potential to reduce overall catastrophic risk from a somewhat different perspective.

The final comment: "PSM is never done. You never finish it."

Have some process in place to sustain and/or continuously improve PSM by using such tools as:

- CPH and SIF initiatives,
- annual site and corporate PSM/health and safety goals,
- PSM conferences, and
- sharing of PSM event learning (both internal and external).

These are key to ensuring the organization keeps the Catastrophic Prevention Journey and PSM alive.

Acknowledgements and attachments

The author would like to thank Lilly management and the reviewers listed below for their support in sharing experiences and documents in this case study. I would also like to thank the ladies and gentlemen whose passion and hard work successfully implemented this vital program at Lilly. It truly has been a privilege to share this Catastrophic Prevention Journey with you.

Reviewers: Steve Gillman (retired), Allan Holmberg (retired), Judith Macklin, Dr. Bernard McGarvey, Brian Reising, and William Walasinski

Items provided in the files on the Web accompanying this book:

- Process Risk Screening Tool
- PSSR Process Description Document

- PSSR Map
- PSSR Requirements Document
- Safety Critical Operations (SCO) Selection Tool
- PSI Template
- Change Management Response and IAT Tool Comments
- MI List – example
- MITI Guideline
- CPH Risk Screening Tool and Worked Example
- CPH Program
- Serious Injury or Fatality (SIF) Pre-Cursor Guide

APPENDIX III: RISK BASED PROCESS SAFETY (RBPS) IMPLEMENTATION TOOLS

Implementation is the tough, unheralded "ground game" (to borrow a football/soccer/rugby analogy) of process safety. It can be considered to consist of the following steps, as encapsulated by the acronym BPTA:

B = Business processes for critical work integrated into existing processes that are supported by line management and sponsored by upper management

P = Procedures and/or tools based on and supporting business processes that can be understood and/or followed by both those executing the business processes and those impacted by them

T = Training and/or communication of the procedures and/or tools, including verification of understanding

A = Auditing and/or self-assessments to ensure sustainment and lack of decay in all three steps above

Perhaps the least understood concept included in these steps is the "RBPS Implementation Tool." In addition, this concept (or rather the underutilization and lack of support for it) is the root cause of many failed RBPS implementation attempts. A tool can take many forms, from a simplistic, inexpensive item to a sophisticated, multimillion-dollar item. It may include any or all of the following:

- A template/form
- An Excel® spreadsheet
- A complex written guideline
- A sophisticated IT database

The following table is based on the RBPS elements from *Guidelines for Risk Based Process Safety* (CCPS, 2007). It is an attempt to assist the reader in determining what tools may be needed when implementing an RBPS or similar PSM management system.

RBPS Element	Tool(s)	Purpose of Tool	Example of Tool
Commitment to Process Safety			
1. Process Safety Culture	Process Safety Culture Survey	To initially assess (and periodically re-assess) Process Safety Culture with regard to Key Principles and develop action plans to improve	Appendix G from *The Report of The BP U.S. Refineries Independent Safety Review Panel* (2007) (A copy is provided in the files on the Web accompanying this book)
	Periodic Self Assessments/ Metrics	To monitor Process Safety Culture	Section 8 (Tier 4 Performance Indicators—Operating Discipline & Management System Performance) from ANSI/API RP 754 (*Process Safety Performance Indicators for the Refining and Petrochemical Industries*) (Available at (http://publications.api.org/)
2. Compliance with Standards	PSM Applicability or Process/ Risk Screening	To identify covered Process Safety Processes and to document their boundaries	Process Risk Screening Tool (See Lilly Item #1 in the files on the Web accompanying this book)
	List/ Guideline /Corporate Standard(s) of relevant process safety standards, codes, regulations and laws, over the life of the process, organized for Process Safety Management (PSM) covered facilities and specific equipment	- To ensure the safe design, installation, inspection, and testing of facilities and equipment occurs - To comply with Recognized and Generally Accepted Good Engineering Practices (RAGAGEP)	Mechanical Integrity Test and Inspection Guideline [See Lilly Item #9 (MITI Guideline) in the files on the Web accompanying this book]
3. Process Safety Competency	Awareness Training for employees operating, maintaining and supporting PSM Covered Processes	To communicate PSM covered processes, boundaries, primary hazards and safeguards to personnel	U.S. Chemical Safety Board incident reports and videos – see www.CSB.gov *Recognizing Catastrophic Incident Warning Signs in the Process Industries* (CCPS, 2011)
	A short-term and long-term (with a 3-5 year outlook) learning/ communication plan/ strategy	To identify, organize and fund activities (based on business case objectives) that support progress toward organizational PSM objectives and promote PSM competency	RBPS book, Chapter 5

RBPS Element	Tool(s)	Purpose of Tool	Example of Tool
	The framework of process safety knowledge and expertise versus the desired competency level in a "super-matrix" format	Targeted at multiple audiences, ranging from front line chemical operators, mechanics and instrument technicians through senior management, including financial and business executives. Gaps between existing and desired training levels can be identified and potential remedies suggested	End product of Project 239: *Guidelines for Process Safety Knowledge and Expertise*
	A process to share PSM incidents and learning, both from internal and external PSM incidents	To develop action plans from learning as appropriate and to maintain a sense of vulnerability	- CCPS Process Safety Beacon – see www.aiche/CCPS - Similar company internal incident sharing documents
4. Workforce Involvement	PSM specific Employee Participation Plan	- To document and communicate PSM related workforce involvement opportunities - To comply with government regulations – if applicable	Example Employee Participation Plan (A copy is provided in the files on the Web accompanying this book)
5. Stakeholder Outreach	Hazard Assessment Study and/or Facility Siting Study	To identify areas of concern in worst-case and alternative release scenarios for PSM covered processes	*Guidelines for Evaluating Process Plant Buildings for External Explosions, Fires, and Toxic Releases*, 2nd Edition (CCPS, 2012)
	Stakeholder Outreach Plan	- To identify offsite and onsite organizations, people, facilities and processes that may be affected - To document how to communicate hazards, safeguards and to both develop and communicate emergency response plans for releases	RBPS book, Chapter 7

RBPS Element	Tool(s)	Purpose of Tool	Example of Tool
Understand Hazards and Risks			
6. Process Knowledge Management	List of Process Safety Information for each covered PSM process, that covers: - Hazards of the materials - Hazards of the equipment - Hazards of the process	- To ensure Process Engineers understand what types of Process Safety Information are needed and the scope/ complexity of the information - To ensure appropriate information is considered for Process Hazard Reviews and Changes - To ensure regulatory requirements for Process Safety Information are met	Process Safety Information (PSI) Template (See Lilly Item #1 in the files on the Web accompanying this book) *Guidelines for Process Safety Documentation* (CCPS, 1995)
7. Hazard Identification and Risk Analysis	Procedure to detail the process to follow for Hazard Identification/ Risk Analysis	- To identify the appropriate Hazard Identification/ Risk Analysis tools for processes - To ensure the appropriate personnel participate - To ensure the appropriate personnel review and approve recommendations and action plans - To ensure that recommendation action plans are resourced and tracked to resolution	*Guidelines for Hazard Evaluation Procedures, 3rd edition* (CCPS, 2008)

RBPS Element	Tool(s)	Purpose of Tool	Example of Tool
	Software tools for documenting hazard studies, such as HAZOPs, LOPAs, FMEAs, and associated checklists	- To allow quicker, more efficient, and consistent completion of PHA worksheets, etc. - To enable better team review/consensus by projecting the tables/checklists - To allow easy incorporation of risk ranking, LOPA calculations, etc.	See the "PSM Software Compilation" in the files on the Web accompanying this book
Manage Risk			
8. Operating Procedures	Procedure to detail the process to follow in developing Operating Procedures	To ensure process owners include the following in procedures: - Operating phases (i.e. shutdown, startup, normal operations, emergency operations, etc.) - Safe Operating Limits - Hazard information - Safety systems and their functions	*Guidelines for Writing Effective Operating and Maintenance Procedures* (CCPS, 1996)
9. Safe Work Practices	Safe Work Procedures/Permits for highly hazardous tasks/ locations, including the following: - Hot Work - Confined Space Entry - Lockout/Tagout - Line Breaking - Access Control	To ensure appropriate risk assessment and safety precautions are taken for highly hazardous tasks/locations	RBPS book, Chapter 11

RBPS Element	Tool(s)	Purpose of Tool	Example of Tool
10. Asset Integrity and Reliability	A List/ spreadsheet that logs all Process Safety covered equipment, facilities and their installation, maintenance and inspection types and frequencies, as well as the regulatory standards/ practices they are based on	To ensure Process Safety covered equipment is identified, along with their safety function, test and inspection frequencies and types of test and inspections	Mechanical Integrity List (See Lilly Item #8 in the files on the Web accompanying this book)
	A guideline for Mechanical Integrity Tests and Inspections	To ensure that tests and inspections are based on appropriate Recognized and Generally Accepted Good Engineering Practices (RAGAGEP – i.e., Codes and Standards)	Mechanical Integrity Test and Inspection Guideline [See Lilly Item #9 (MITI Guideline) in the files on the Web accompanying this book] *Guidelines for Mechanical Integrity Systems* (CCPS, 2006)
	Software tools for managing maintenance and/or the overall Mechanical Integrity program	To ensure tests are performed on schedule, adequately documented, and results reviewed/analyzed	See the "PSM Software Compilation" in the files on the Web accompanying this book
11. Contractor Management	A checklist/process that identifies Contractor HSE hazards/aspects	To ensure process safety hazards/chemicals are identified and their associated safeguards before work is undertaken	*Recommended Guidelines for Contractor Safety and Health* (Texas Chemical Council, 2008) (A copy is provided in the files on the Web accompanying this book) *Contractor and Client Relations to Assure Process Safety* (CCPS, 1996)
	Hazardous process overview training and safety requirements	To ensure knowledge of hazardous materials and processes are conveyed to the contractor	*Recommended Guidelines for Contractor Safety and Health* (Texas Chemical Council, 2008) (A copy is provided in the files on the Web accompanying this book)
	Permits/process to communicate local/task hazards and required safeguards	To ensure local hazards/ hazards of planned tasks are communicated, as well as chemicals brought in by contactors and the associated safeguards to be employed	RBPS book, Chapter 13

RBPS Element	Tool(s)	Purpose of Tool	Example of Tool
12. Training and Performance Assurance	A checklist/ template (often referred to as a Training Matrix) for Training Verification	To ensure that required training is identified for each job, completed on schedule, and adequately completed/documented	Example training workflow and matrices
13. Management of Change (MOC)	MOC and/or Corrective Action/ Preventative Action (CAPA) tool (with processes to create a unique identifier and assign a change level based on scope/ complexity of the change) for each change. This can be either: - A simple paper-based form accompanying the hard copy information for the change (with the unique identifier and change level logged onto a separate tracking spreadsheet) - A sophisticated database/ IT driven process that requires electronic acknowledgement/ signatures for the change to proceed	- To ensure appropriate resources are allocated for the scope of the change - To ensure an orderly process is followed for those assessing/ reviewing and approving the change - To ensures changes are not "lost" - To ensure that important implementation steps/ action plans are not missed, including hazard reviews, procedures, training, Pre-Startup Safety reviews, etc. - To ensure regulatory requirements for Process Safety and other business regulations, such as Quality (e.g. Food and Drug Administration requirements) are met	Numerous IT databases are on the market. An example of a database used by many pharmaceutical firms is the TrackWise (registered trademark) database See the "PSM Software Compilation" in the files on the Web accompanying this book

RBPS Element	Tool(s)	Purpose of Tool	Example of Tool
	A hard copy or electronic set of Safety and Health driven questions that identifies the major hazards and the scope of those hazards to those executing the change	- To ensure the appropriate resources are allocated to the change in terms of H&S reviews/ assessments - To ensure that important implementation steps/ action plans are not missed, including hazard reviews, procedures, training, Pre-Startup Safety reviews, etc.	Change Management Response and Impact Assessment Task Comments - see Appendix in Lilly Case Study *Guidelines for the Management of Change for Process Safety* (CCPS, 2008)
	An integrated or separate process to manage specified organizational changes	To ensure that organizational changes do not negatively impact the PSM system or its performance	*Guidelines for Managing Process Safety Risks During Organizational Change* (CCPS, 2013)
14. Operational Readiness	Pre-Startup Safety Review Process	A process review/ timeout to ensure critical steps are taken that will prevent Process Safety incidents, prior to hazardous chemicals being introduced to a process. This includes startups after a: - Project - Operational Shutdown - Emergency Shutdown - Maintenance Shutdown	PSSR Process Description and Process Map (See Lilly Items #2 and #3 in the files on the Web accompanying this book)
	Pre-Startup Safety Review Checklists/Procedures(s)	Ensures critical pieces of equipment (e.g. valves aligned), procedures, training, maintenance and Process Safety Information are updated prior to start-up	See Lilly Item #4 (PSSR Requirements Document) in the files on the Web accompanying this book *Guidelines for Performing Effective Pre-Startup Safety Reviews* (CCPS, 2007) (PSSR checklists and procedures are included in the CD with that book)

RBPS Element	Tool(s)	Purpose of Tool	Example of Tool
15. Conduct of Operations	Tools for managing the People, Process, and Plant aspects of a facility in a consistent, disciplined way	To ensure reliable, consistent, and correct execution of the policies, procedures, and practices that make up a facility's risk management system	*Conduct of Operations and Operational Discipline: For Improving Process Safety in Industry* (CCPS, 2011)
16. Emergency Management	Plans and processes for thorough planning, effective training, realistic drills, effective two-way communication with stakeholders, and adherence to plans/procedures – in advance of an incident	To ensure effective, tested emergency response plans, trained and equipped response teams, and effective methods of protecting personnel, the environment, and property	*Guidelines for Technical Planning for On-Site Emergencies* (CCPS, 1995) RBPS book, Chapter 18
Learn from Experience			
17. Incident Investigation	Incident Investigation Forms	To ensure consistent required information is gathered and communicated for Incidents	*Guidelines for Investigating Chemical Process Incidents*, 2nd Edition (CCPS, 2003)
	Root Cause Analysis Methodology	To ensure formal investigations are conducted consistently and accurately	*Guidelines for Investigating Chemical Process Incidents*, 2nd Edition (CCPS, 2003)
	Software tools for reporting, investigating, and approving incident reports; performing root cause analyses; and for managing the corrective actions	To enable consistent and complete reporting, thorough documentation, and provide comprehensive management of corrective actions	See the "PSM Software Compilation" in the files on the Web accompanying this book

RBPS Element	Tool(s)	Purpose of Tool	Example of Tool
18. Measurement and Metrics	Metrics that are (1) sensitive enough to help facility management monitor the performance and efficiency of the RBPS management system on a near-real-time basis and (2) communicated to the appropriate personnel for routine monitoring (or as "bulletins" when there are significant or abrupt changes)	To identify evolving management system weaknesses and make adjustments to RBPS element work activities before they degrade into a "failed" state (in terms of performance or efficiency)	ANSI/API RP 754 (*Process Safety Performance Indicators for the Refining and Petrochemical Industries*) (Available at (http://publications.api.org/)
19. Auditing	Auditing and/or Self-Assessment protocols	To ensure requirements are consistently and thoroughly assessed or audited	*Guidelines for Auditing Process Safety Management Systems*, 2nd Edition (CCPS, 2011)
	Audit protocols/checklists covering various aspects of a facility, its processes, and its PSM and HSE management systems	To identify significant risks and potential regulatory deficiencies prior to an acquisition To ensure that potential gaps between the old and new company's standards are recognized and addressed during integration	*Guidelines for Acquisition Evaluation and Post Merger Integration* (CCPS, 2010)
20. Management Review and Continuous Improvement	A process in which the management review team meets regularly with the individual(s) responsible for each RBPS element to (1) present program documentation and implementation records, (2) offer direct observations of conditions and activities, and (3) answer questions about program activities	To provide regular checkups on the health of PSM systems in order to identify and correct any current or incipient deficiencies before they might be revealed by an audit or incident	RBPS book, Chapter 22

APPENDIX IV: THE BUSINESS CASE FOR PROCESS SAFETY

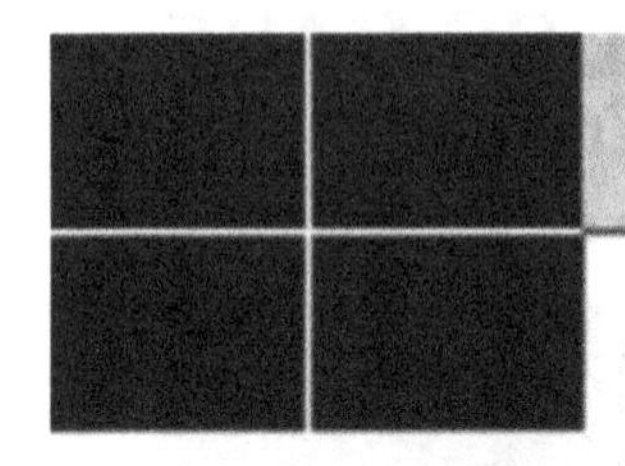

THE CENTER FOR
CHEMICAL PROCESS SAFETY

SINCE 1985, The Center for Chemical Process Safety (CCPS) has helped manufacturers, insurers, government, academia, and consultants work together to improve manufacturing process safety.

CCPS and its sponsors are committed to protecting employees, communities, and the environment by developing engineering and management practices to prevent or mitigate catastrophic releases of chemicals, hydrocarbons, and other hazardous materials.

CENTER FOR CHEMICAL PROCESS SAFETY
AMERICAN INSTITUTE OF CHEMICAL ENGINEERS
3 Park Ave, New York, N.Y., 10016-5991, U.S.A.
Tel: (212) 591-7319 • Fax: (212) 591-8883
E-mail: ccps@aiche.org • www.ccpsonline.org

Library of Congress Cataloging-in-Publication Data

The business case for process safety : corporate responsibility, business flexibility, risk reduction, sustained value / The Center for Chemical Process Safety.
p. cm.
[illegible]

[illegible]

THE BUSINESS CASE FOR PROCESS SAFETY

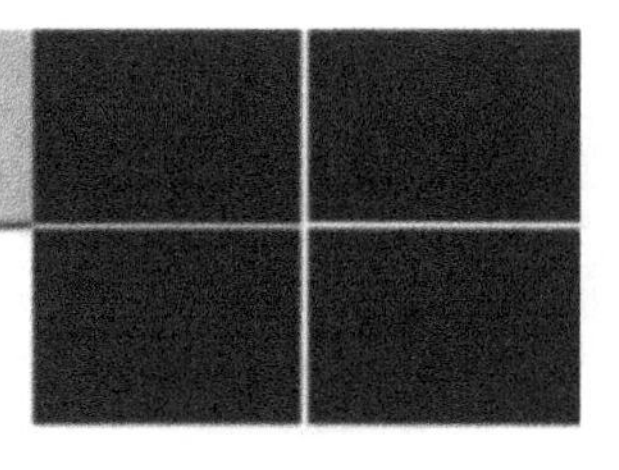

WHAT SEPARATES THE TOP-PERFORMING COMPANIES IN OUR INDUSTRY FROM THE REST?

One essential characteristic CCPS member companies display is that they have adopted a rigorous philosophy regarding process safety. This summary of a recent industry-wide study identifies four ways that your business will benefit from implementing a robust process safety program. Process safety is an essential part of achieving manufacturing excellence and increasing profitability and shareholder value no matter the size of the enterprise.

We have seen process safety benefit our business in ways we had not anticipated. We need to share this message with others.

ARNOLD ALLEMANG
Vice President – Operations
The Dow Chemical Company

Corporate Responsibility

Risk Reduction

Business Flexibility

Sustained Value

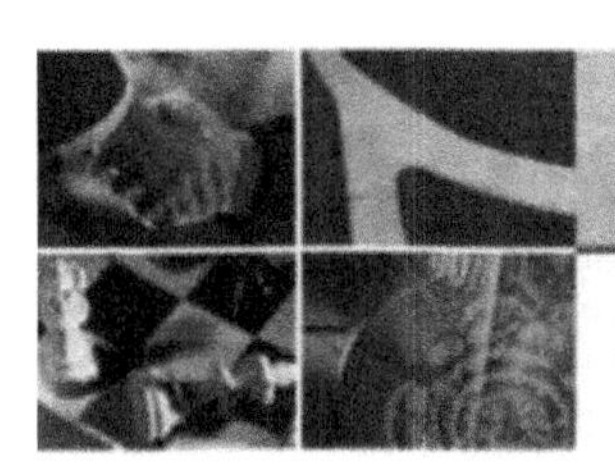

THE BENEFITS OF PROCESS SAFETY

THIS BENCHMARK STUDY of CCPS member companies, combined with data from other sources, provides conclusive evidence that methodically implementing process safety provides four benefits essential to any healthy business. Two of these benefits are qualitative and as a result are somewhat subjective. You can see them in the way the public, your shareholders, government bodies, and your customers relate to your company. The two remaining benefits are quantitative. These have measurable impact in terms of your bottom line and company performance. All four benefits, when realized together by adhering to a sound process safety system, combine to support the profitability, safety performance, quality, and environmental responsibility of your business.

TWO QUALITATIVE BENEFITS

Corporate Responsibility

Process safety helps your company display corporate responsibility through its actions. The heart of process safety lies in consistently planning to do the right things, then doing them right — consistently. Corporate responsibility leads to the second benefit…

Business Flexibility

Corporate responsibility as demonstrated in your process safety management program leads to a greater range of business flexibility. When you openly display responsibility through implementing an effective process safety program, your company can achieve greater freedom and self-determination.

TWO QUANTITATIVE BENEFITS

Risk Reduction

A healthy process safety program significantly reduces the risk of catastrophic events and helps prevent the likelihood of human injury, environmental damage, and associated costs that arise from incidents. Although the essence of process safety focuses on preventing catastrophic incidents, the number of less severe incidents is also reduced.

Sustained Value

Process safety relates directly to enhanced shareholder value. When properly implemented, it helps ensure reliable processes that can produce high quality products, on time, and at lower cost. These improvements allow the enterprises that make them to sustain value creation over time.

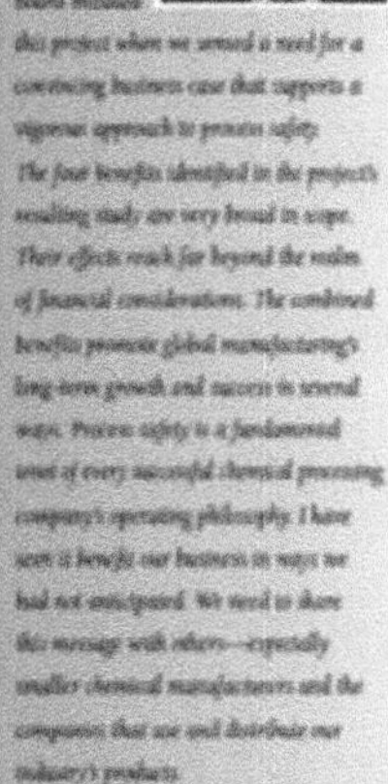

My colleagues and I on the CCPS advisory board initiated this project when we sensed a need for a convincing business case that supports a vigorous approach to process safety. The four benefits identified in the project's resulting study are very broad in scope. Their effects reach far beyond the realm of financial considerations. The combined benefits promote global manufacturing's long-term growth and success in several ways. Process safety is a fundamental tenet of every successful chemical processing company's operating philosophy. I have seen it benefit our business in ways we had not anticipated. We need to share this message with others—especially smaller chemical manufacturers and the companies that use and distribute our industry's products.

ARNOLD ALLEMANG
Vice President – Operations
The Dow Chemical Company

4

WHAT IS PROCESS SAFETY?

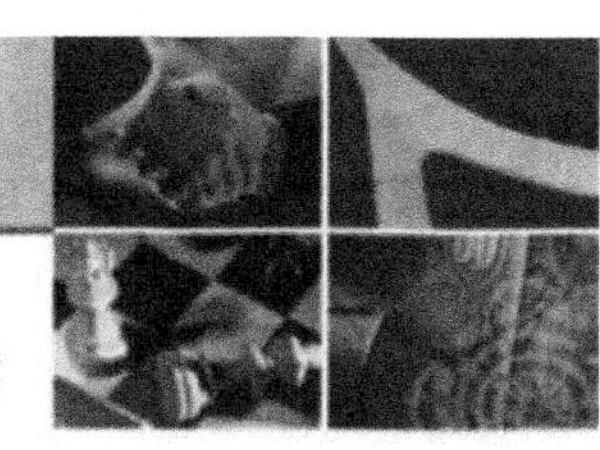

PROCESS SAFETY is a management system implemented to prevent major incidents involving hazardous materials. It is necessary for managing complex chemical operations. A process safety management system focuses on three important aspects of your business:

TECHNOLOGY

This component includes developing accurate process safety information about your equipment and technology, performing process hazard analyses, developing operating procedures and safe work practices, and then managing changes as they arise. It also includes designing manufacturing processes that are inherently safer from their conception.

FACILITIES

This aspect focuses on the mechanical integrity of your plant's equipment and the software that controls it. It includes preventive maintenance programs, performing pre-startup safety reviews, and aligns with management of change to help ensure continuous safe operation. Good design and maintenance along with periodic safety reviews protect your company's means of production.

PERSONNEL

Involving your employees in building and maintaining your process safety program is the best way to communicate its ongoing importance throughout the organization. Other process safety elements involving personnel include training employees in process hazards and their job tasks, managing contractors properly, investigating incidents to understand their root causes, implementing actions to prevent recurrence, preparing for emergencies, planning effective response, and self-auditing to gauge performance and to identify opportunities for improvement in all three aspects of process safety.

Understanding the skills and knowledge required for a job when making changes in work assignments will help reduce errors and improve safe performance.

The personnel aspect of process safety leads companies to minimize turnover of key personnel at all levels and maximize corporate memory of experiences, best practices, and industry lessons.

The nature of our business is one that requires a high level of managerial, technical and operational discipline. The discipline practiced when implementing and maintaining a healthy process safety program easily translates to other business areas and helps address other business risks.

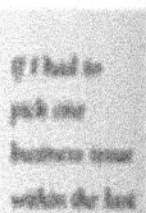

If I had to pick one business issue within the last 12 years that has benefited our company most, I would say it was the advent of process safety. It not only helps prevent incidents and reduce their consequences, it also provides fundamental methods for managing our business. It gives us valuable guidance for operating our health, environmental, and safety programs more effectively and provides structured systems, such as management of change, which we apply in all aspects of our business.

STEVE KEMP
Vice President – Health, Environment and Safety
Occidental Chemical Corporation

5

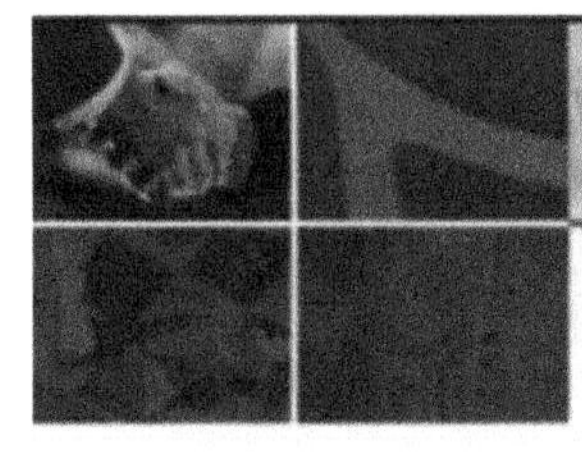

CORPORATE RESPONSIBILITY

COMMUNITIES NEAR your production sites respect companies that care about their employees and the people living nearby, and that contribute to the local government and economy. Implementing a strong process safety program helps your company display the following attributes in a way that will help neighbors and employees understand your commitment to being a responsible neighbor. Proactively managing an effective process safety program displays a high level of corporate responsibility and encourages you to sustain it long-term. It helps with:

- Fulfilling your obligation to protect employees and the community
- Enhancing customer and supplier relationships
- Complying with regulations
- Conforming to industry standards

Committing to a dynamic process safety program displays corporate responsibility and social responsibility

Displaying responsibility helps a company in many ways:

- It helps your investors perceive a lower risk when they make buying or selling decisions.
- It is the best insurance policy for protecting company reputation and shareholder value.
- It increases the value of your corporate image and brand.
- It reduces concerns within the local community.
- It engages employees at all levels by increasing morale, loyalty, and retention.
- It improves your ability to get insurance coverage at attractive rates.
- It enhances your lenders' confidence thus helping with capital expansion.
- It helps regulators understand your facility's credibility and unique considerations.

Companies of all sizes can benefit by establishing a positive reputation for leadership in the community.

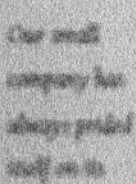

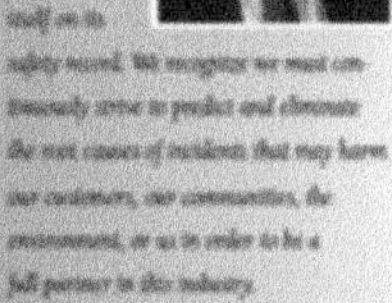

Our small company has always prided itself on its safety record. We recognize we must continuously strive to predict and eliminate the root causes of incidents that may harm our customers, our communities, the environment, or us in order to be a full partner in this industry.

TOM REILLY
Chairman of the Board
Reilly Industries, Inc.

6

BUSINESS FLEXIBILITY

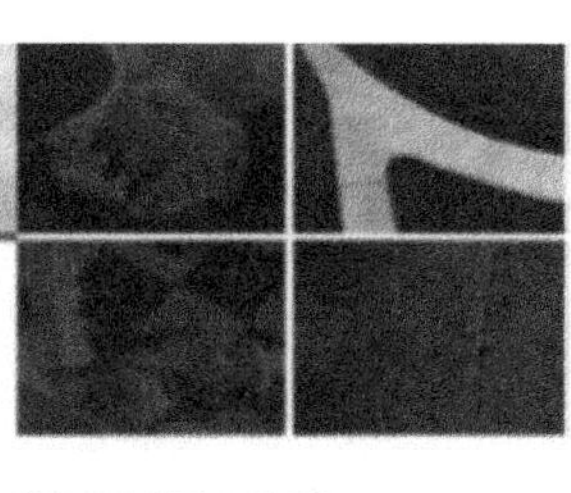

COMPANIES that manage process safety effectively earn the flexibility to freely manage their businesses and grow profitably, while satisfying all stakeholders — local communities, the public, regulatory authorities, governments, investors, and customers. This business flexibility, or self-determination, is a result of the trust the public and the local community have in you. It is your corporation's license to operate. This analogy is complete in the sense that, as with any license, those who award it can also revoke it.

Any major industrial incident can result in the creation of more prescriptive regulations that will affect the entire industry. After a major incident, every company in a related business suffers a loss of public trust, which may result in increased regulation and compliance costs. The livelihood of every employee of your company — from officers and board members to managers, technical staff, and hourly workers — is on the line when you operate without an effective process safety program.

A company's future existence is at stake after it experiences a major incident at one of its sites. Managing a rigorous process safety management system helps maintain a company's freedom and self-determination, allows innovation, and ensures a greater range of business flexibility. A company's freedom to operate can be severely compromised due to community discontent, regulatory scrutiny, legal complications, and even intervention by a company's own board of directors when key stakeholders sense increased risk.

This flexibility benefits a company by:

- Proving your worthiness to hold a license to operate
- Strengthening and maintaining good relationships with the local community
- Lowering interest rates for financing
- Helping you attract and retain high performance staff
- Helping you obtain approvals for expansion permits or new facilities more quickly — a critical strength when manufacturing new products needed to compete effectively
- Allowing managers to focus on sales and growth, rather than the last accident
- Strengthening and maintaining good relationships with regulators

Demonstrating a strong process safety culture within all levels of your organization gives your company a greater degree of business flexibility.

Safety, including process safety, is a guiding principle for our company. We believe that the traits required to achieve excellence in safety are the same as those required to achieve outstanding results in all other aspects of our business. Safety is simply good business.

RALPH F. HERBERT
Vice President, Engineering
ExxonMobil Chemical Company

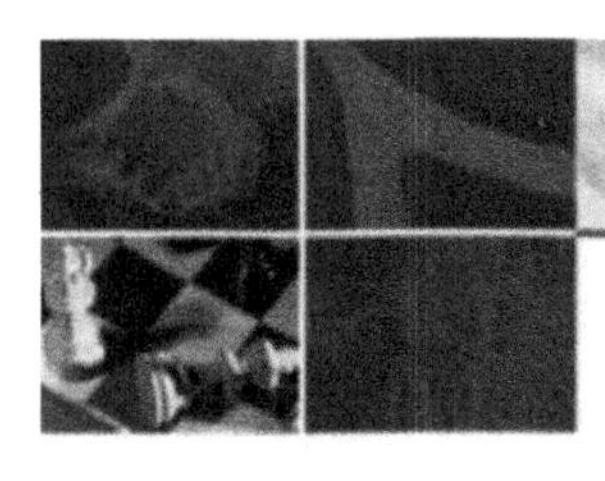

RISK REDUCTION

A robust process safety program will help your company reduce risk and avoid loss.

COMPANIES that implement effective process safety programs receive a windfall benefit every year. This income shows up on the bottom line in the form of the major incident that DID NOT occur. Process safety provides unparalleled capacity for enhanced risk reduction. Your company's risk exposure is reduced in the following areas when well-founded process safety systems are in place.

- **Lives are saved and injuries are reduced** — Both the personal impact of human loss and cost of deaths or injuries are painful. A solid process safety program can help prevent these costs.
- **Property damage costs are reduced** — In the U.S., major industrial incidents cost an average of $80 million each.
- **Business interruptions are reduced** — These losses can amount to four times the cost of the property damage from an incident.
- **Loss of market share is reduced** — After an incident, this loss continues until the company's reputation is restored. Adverse publicity and negative public image can have insurmountable effects.
- **Litigation costs are reduced** — These are unavoidable after an incident and can total five times the cost of the regulatory fines.
- **Incident investigation costs are reduced** — Investigating an incident and implementing corrective actions can cost millions of dollars.
- **Regulatory penalties are reduced** — For many incidents, a fine after litigation can total 1 million dollars or more.
- **Regulatory attention is reduced** — A major incident usually results in increased regulatory audits and inspections.

Any of these items can easily put a smaller company out of business.

How does increased risk reduction capability enhance your company?

Most companies participating in this study observed significant reductions in injury rates due to implementing high quality process safety programs.

- One company achieved a 50% reduction in injuries and fatalities resulting from major incidents (compared to overall industry averages). This saved them over $5 million per year and an additional savings of $3 million per year in reduced worker compensation costs.
- Incidents cause operational interruptions. If your plant is not idle due to an incident, you are making product and the business can flourish.
- Incidents divert corporate management's attention from long-term business planning. When these managers have to stop and deal with incidents and other crises, it can distract their ability to concentrate on sustainable growth.
- Your corporate reputation and legacy are protected for future generations.

Implementing an exacting process safety program has helped our company maintain the business flexibility we need to meet the challenges ahead. It has helped us to be free to innovate and grow as needed. The good relations we are able to nurture with neighbors and regulators are essential to this flexibility. These relationships depend also upon our ability to ensure that a viable process safety program is in place at all our plants worldwide.

JOACHIM KRUEGER
Vice President Global Environmental, Health and Safety
Celanese AG

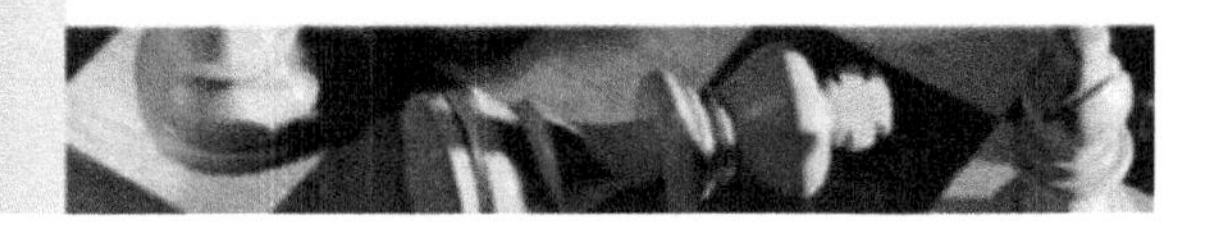

Effective process safety programs provide industry-wide benefits.

Fatalities, injuries, property damage and business interruption can cause a substantial drop in share price and loss of market share for your company. Disruption of normal business activity can cause a temporary loss of corporate direction by diverting senior management's attention from running the business to overseeing damage control. Additionally, company officers may be subject to personal liability and even criminal charges.

After an incident, a ripple effect can occur throughout a large company and can ultimately traverse the entire industry.

- What happens at your plant in Baton Rouge can affect your plants in Brussels and Beijing.
- An incident at a small, unrelated company can negatively affect the public's perception of much larger, well-managed companies.
- An incident at a key raw material supplier's facility can keep you from meeting obligations to your customers and possibly affect an entire business area.

A major incident can place a company in a position where it is unable to respond to its competitors' business actions. A company in a weakened state may become subject to an undesirable takeover. This situation reverberates throughout the industry.

Material Damage and Business Interruption Costs from Incidents

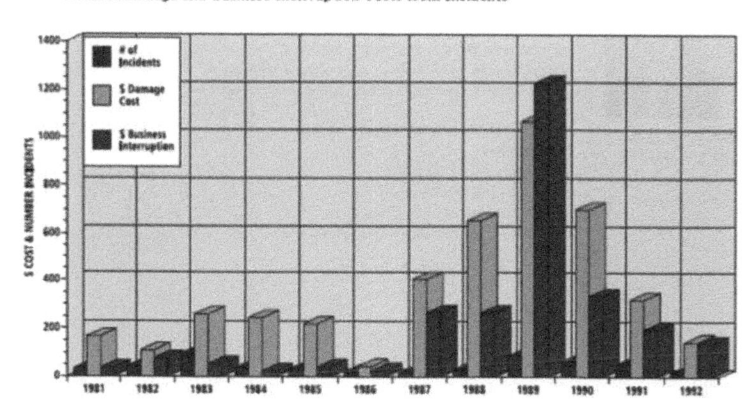

Source: MIACC (1996) based on data from Swiss Reinsurance Company, Zurich, Switzerland

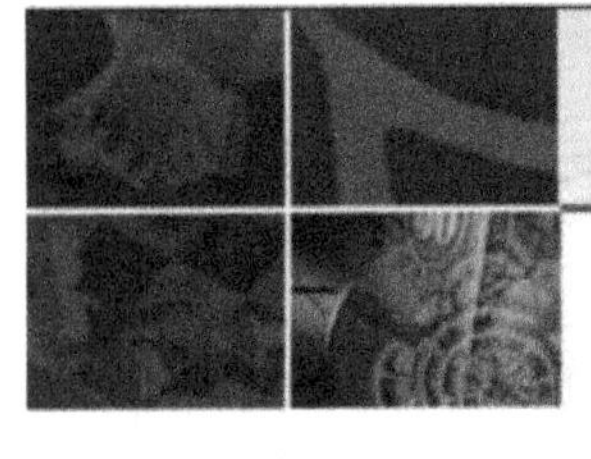

CREATING SUSTAINED VALUE

Implementing an effective process safety program will create and sustain value for your company and its shareholders. This is a bonus with additional benefits.

Embracing process safety as an essential part of the way you do business allows your company to achieve a measurable increase in revenues and a reduction in costs. Creating value is a complement to reducing risk and avoiding loss. The value created can be substantial. In some cases, product stewardship requirements make a strong process safety program a necessity for your company to do business. The chemical processing and petroleum companies that participated in this study report the following returns from their investment in process safety:

- **Productivity Increases** — Up to 5% increases in productivity, due mainly to increased reliability of equipment. Increased revenues of $50 million are reported.
- **Production Costs Decrease** — Up to a 3% reduction in production costs resulting in a savings of $30 million.
- **Maintenance Costs Decrease** — Up to a 5% reduction in maintenance costs resulting in savings of $50 million.
- **Lower Capital Budget Required** — Up to a 1% reduction in capital budget resulting in savings of $12 million.
- **Lower Insurance Premiums** — Up to a 20% reduction in insurance costs resulting in savings of $6 million.

Process safety helps to increase productivity through:

- Improved reliability and mechanical integrity of equipment, causing fewer operational interruptions
- User-friendly, accurate operating procedures and safe work practices
- Improved team effectiveness through effective employee training programs
- Employee ownership of the systems to help ensure their safety and the safety of the community
- Enhanced troubleshooting capabilities for all types of production issues
- Identifying and addressing safety, operability, and reliability issues before they occur
- Decreased turnaround time for minor repairs or replacement of equipment
- Extended intervals between major turnarounds and reduced turnaround time

Production costs are reduced through:

- Improved yields
- Lower costs to rework off materials
- Lower costs for waste stream disposal
- More efficient staff requiring less supervision
- Engaged employees participating in continuous improvement

At 3M, we have found the benefits of process safety are the obvious ones — a safer workplace, business continuity assurance, and improved employee morale. In fact, we apply process safety to non-regulated processes as a best management practice.

RONALD R. BELSCHNER
Vice President,
Engineering, Manufacturing & Logistics
3M Company

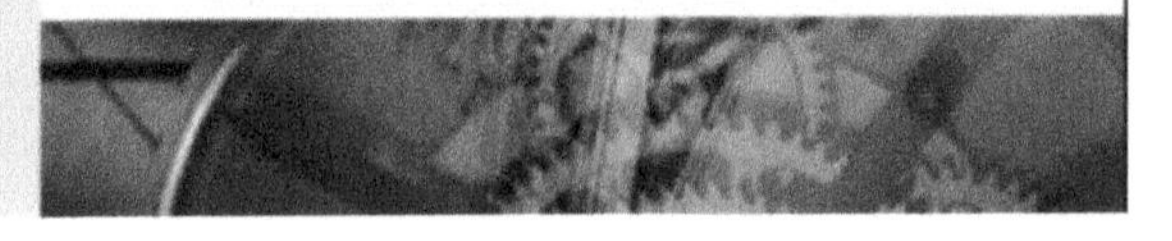

Maintenance costs are reduced through:

- Effective written equipment maintenance procedures
- Contractor safety programs
- Repairing or replacing critical equipment before it fails
- Lower maintenance turnaround costs

Capital budgets are reduced through:

- Process hazard analyses for new projects and facilities to mitigate risk over their useful life
- Inherently safer process designs that begin in the conceptual phase
- Lower capital expenditures because project teams have up-to-date process safety information

Insurance costs are reduced through:

- Effective emergency planning and response which helps to reduce the loss from an incident and helps prevent minor incidents from developing into a major incident
- Thorough incident reporting and investigation programs to prevent incidents from being repeated
- Reporting and investigation of "near misses" to identify potential problems early
- Lower casualty insurance premiums when your insurers detect that an effective process safety program is in place to lower the probability of major incidents

Sustained Value Data

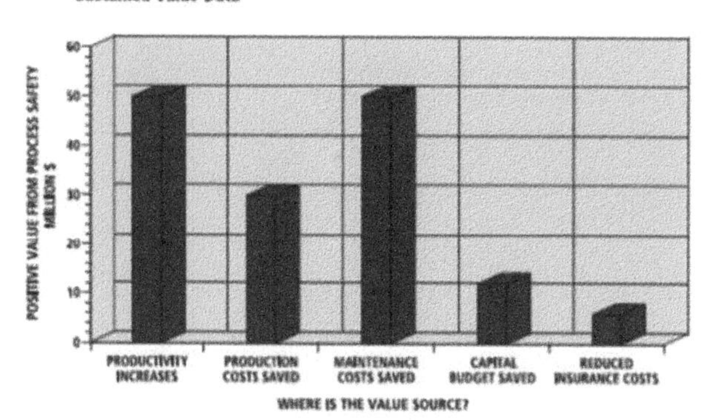

CCPS Study: 1998 Workshops

WHAT PROCESS SAFETY CAN DO FOR YOU

YOUR PROCESS SAFETY PROGRAM also supports other programs. It supports quality, environmental responsibility, industrial hygiene, worker safety, and sustainable development.

Your staff has probably detected the natural synergy between process safety and other business requirements you may face. Implementing a process safety program provides a management system model that can be modified for implementing other programs focused on:

- **Occupational Safety Requirements** — Both internal and governmental
- **Quality Management** — Customer driven quality systems (including, ISO, FDA and others)
- **Environmental Requirements** — Internal, governmental, or customer driven
- **Profitability** — Internal business management systems to help ensure a sustainable, healthy business

Think about it — one common management system model applied throughout an organization can provide a framework for developing the work processes, procedures, and documentation you need for all your safety, quality, environmental, and business commitments. Too often, companies implement a program to comply with one regulation or industry standard and find that another standard or regulation requires a similar element; such an example is management of change. Adopting process safety as the structure upon which to model your integrated management systems will help avoid redundancy and allow productive direction of your company's energies. This logically leads to efficient production and increased shareholder value.

In summary, a robust process safety program will enhance your business in these four ways:

- It displays your company's high level of corporate and social responsibility.
- It allows your company a greater range of business flexibility — the freedom to manage your business.
- It helps your company manage risk and prevent major losses.
- It creates sustained value for your company and its shareholders.

Aside from being the obviously right thing to do from safety and environmental stewardship perspectives, unrelenting commitment to a strong process safety management system is fundamental to a sustainable business model. Undesirable incidents of any sort detract from the value of a business, but a process safety incident has a negative impact on all stakeholders... customers, shareholders, employees, and the communities in which a plant operates. The bottom line is that outstanding process safety performance is a pathway to both financial success and your license to operate.

JAMES B. PORTER
Vice President of Engineering and Operations
DuPont

THE PATH FORWARD

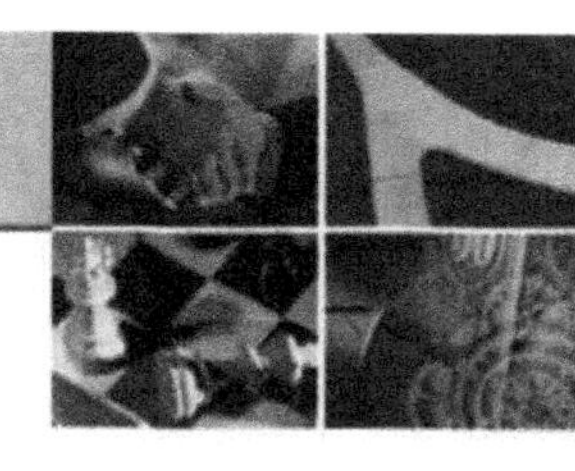

REALIZE THE BENEFITS

Seven steps to achieving business excellence through process safety management:

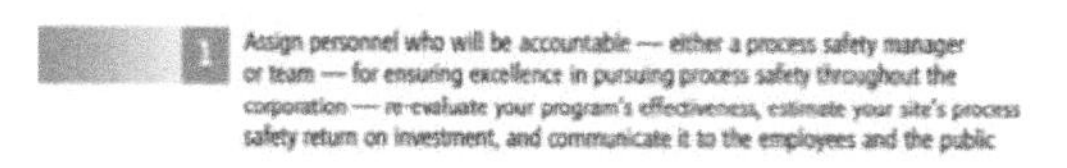

1. Assign personnel who will be accountable — either a process safety manager or team — for ensuring excellence in pursuing process safety throughout the corporation — re-evaluate your program's effectiveness, estimate your site's process safety return on investment, and communicate it to the employees and the public

2. Adopt a personalized company philosophy of process safety. Use it to establish a management system along the lines of CCPS guidelines and tie it to your company's core values

3. Learn more about process safety by reviewing the literature and other references, attending training provided by process safety professionals, and interacting with other companies — networking with them and participating in industry alliances

4. Take advantage of the strong synergy process safety has with your other business drivers — total quality management (TQM), regulatory requirements, and the American Chemistry Council's Responsible Care® initiative all share common elements

5. Set achievable process safety goals that will support the business case presented here over the next one to five years

6. Track your performance versus goals periodically

7. Revisit your process safety program and modify it every three to five years as needed.

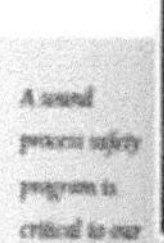

A sound process safety program is critical to our current and continued success. Maintaining it requires discipline in the way we design, operate, and manage our plants. To keep our process safety performance at the high level we demand, we must nurture knowledge levels and leadership qualities throughout each business and at the very top of our organization. Our goal is to ensure each employee understands their role in the safe operation of our chemical sites and the value that brings to the Company. This message extends to our customers, our suppliers, and, in fact, to the entire industry. We simply have no choice but to seek to be leaders in process safety.

TOM ARCHIBALD
Vice President, Director of Operations & Manufacturing
Rohm and Haas Company

13

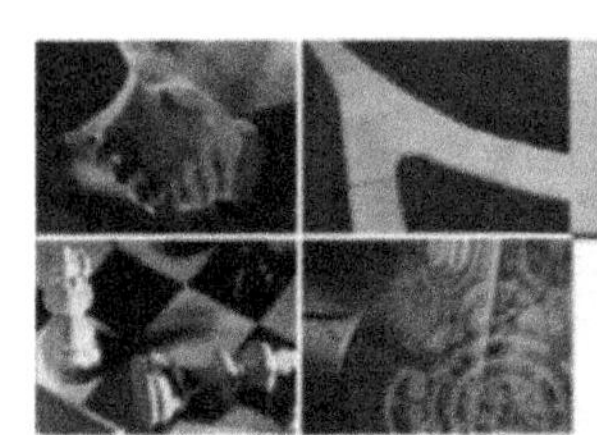

CONTACT CCPS FOR HELP

HOW THE CENTER FOR CHEMICAL PROCESS SAFETY CAN HELP

Contact the Center for Chemical Process Safety to find out about:

- Our process safety guideline series of books and other publications
- The regular networking and conference opportunities that will put you in touch with experienced industry experts
- The effective training resources we can provide — both traditional training courses and computer-based training
- Peer input to help you build or upgrade your process safety program
- How you can participate in projects that will help our industry achieve prime performance in the area of process safety and remain a safe, vibrant, and profitable business
- How your company can become a CCPS member

Visit www.ccpsonline.org for more information or call (212) 591-7319.

Merck's mission is to deliver medicine to the people. Maintaining a dynamic process safety management system throughout our drug development and manufacturing processes allows our business managers to focus on our mission — delivering needed medicines, without interruption, to patients around the world, every day. Just one significant process safety incident could have unimaginable impact on our employees, our communities, our shareholders, our reputation and, most importantly, our patients. Daily attention to process safety management not only prevents these significant losses, it also helps us avoid minor incidents that would disrupt our operations and the medicinal supply chain.

LIAM MURPHY
Vice President, Safety and the Environment
Merck & Company, Inc.

14

THE CENTER FOR
CHEMICAL PROCESS SAFETY

CENTER FOR CHEMICAL PROCESS SAFETY
AMERICAN INSTITUTE OF CHEMICAL ENGINEERS
3 Park Ave, New York, N.Y., 10016-5991, U.S.A.
Tel: (212) 591-7319 • Fax: (212) 591-8883
E-mail: ccps@aiche.org • www.ccpsonline.org

CCPS Book List

Avoiding Static Ignition Hazards in Chemical Operations
Deflagration and Detonation Arrestors
Electrostatic Ignitions of Fires and Explosions
Essential Practices for Managing Chemical Reactivity Hazards
Evaluating Process Safety in the Chemical Industry, Understanding Quantitative Risk Analysis, 2nd edition
Guidelines for Auditing Process Safety Management Systems
Guidelines for Chemical Process Quantitative Risk Analysis, 2nd edition
Guidelines for Design Solutions for Process Equipment Failures
Guidelines for Engineering Design for Process Safety
Guidelines for Evaluating Process Plant Buildings for External Explosions and Fires
Guidelines for Evaluating the Characteristics of Vapor Cloud Explosions, Flash Fires & BLEVEs
Guidelines for Facility Siting and Layout
Guidelines for Fire Protection in Chemical, Petrochemical and Hydrocarbon Chemical Facilities
Guidelines for Hazard Evaluation Procedures (2nd Edition) with Worked Examples
Guidelines for Investigating Chemical Process Incidents, 2nd edition
Guidelines for Mechanical Integrity Systems
Guidelines for Pressure Relief and Effluent Handling Systems
Guidelines for Preventing Human Error in Process Safety
Guidelines for Process Equipment Reliability Data with Data Tables
Guidelines for Process Safety in Outsourced Manufacturing Operations
Guidelines for Safe Automation of Chemical Processes
Inherently Safer Chemical Processes, A Life Cycle Approach
Layer of Protection Analysis: Simplified Process Risk Assessment
Making EHS an Integral Part of Process Design
Plant Guidelines for Technical Management of Chemical Process Safety
Revalidating Process Hazards Analyses
Safe Design and Operation of Process Vents and Emission Control Systems
Understanding Explosions
Wind Flow and Vapor Cloud Dispersion at Industrial & Urban Sites

CCPS would like to thank all
the companies and individuals
who contributed to this study.

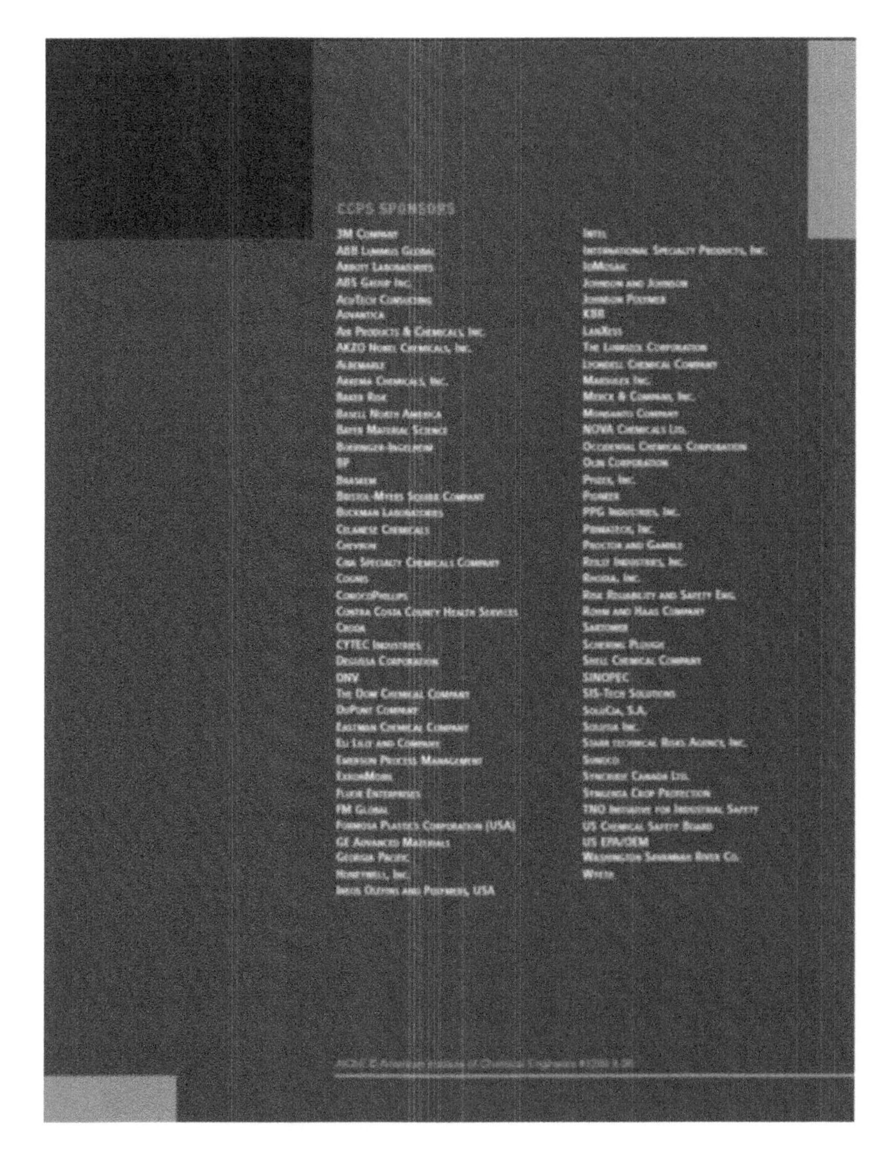
CCPS SPONSORS

APPENDIX V: EXAMPLE FACILITY RANKING PROCESS

(Originally Appendix 5-1 in *Guidelines for Implementing Process Safety Management Systems)*

Notes:

1. The example below is only one of many possible risk ranking approaches, and was copied from the first edition of these guidelines. It is intended to be an initial screening tool to assess overall risk at various company facilities.
2. Other similar approaches commonly used at present include Risk Registers and "Top Risk."
3. Any mention or discussion of risk ranking would be incomplete without mentioning quantitative risk criteria. See *Guidelines for Developing Quantitative Safety Risk Criteria* (CCPS, 2009) for more information on this subject.

1. Collect Data on Quantities and Condition of Hazardous Material

Use the list of hazardous materials contained in OSHA CFR 1910.119, or other similar lists published by industry associations and regulators worldwide. Data on plant sections, quantities of hazardous material, and their storage conditions can be gathered in a number of different ways, depending on what information is available.

- If good process flow diagrams or process and instrumentation diagrams exist they will identify the plant sections.
- If no satisfactory process drawings exist you can identify major sections of the process (e.g., heat exchanger trains, refrigeration circuits, distillation equipment, reactor systems) that probably represent equipment that can be isolated from the rest of the process.
- Data on the quantity and storage conditions of material within the plant sections may be available from a plant material balance.
- The mechanical design drawings for equipment can be used to calculate quantities of material contained in major equipment items. Also include quantities in hold-up within the process area.
- In addition to the quantities in the major equipment items there will be material in the associated pipework and other ancillary equipment. In the absence of better data, add 20 percent to allow for this.

- Estimates of flow rates, residence times, and time required to shut off flows can be used to calculate estimates of potential release quantities. Such information can often be obtained from operating instructions.
- Data on operating conditions can also be obtained from shift logsheets, where critical conditions are often recorded, or by conducting a survey of conditions in the control room and in the field.

2. Estimate Potential Hazard Areas

Different materials pose different hazards, including: thermal radiation, explosion overpressure, and toxic and flammable vapor clouds. Some materials pose only one hazard, while others may pose all four. For the purposes of ranking facilities you will need to estimate the largest area affected by the potential hazards. You can arrive at such an estimate by calculating the greatest downwind distance to a particular level of hazard. The following thresholds are commonly applied:

Thermal radiation 5 kW/m^2 (severe burns to bare skin within 30 seconds)

Explosion overpressure 0.1 barg (minor structural damage, injury from falling masonry, glass, etc.)

Toxic hazard IDLH (immediately dangerous to life and health) or AEGL (acute exposure guideline level) concentration

Flammable vapor cloud Lower flammable limit

Each of these criteria represents a level at which injuries can occur. In addition, some other criteria may be appropriate, for example where existing hazard calculations are in use, corporate policy defines different standards, or local regulation sets criteria.

In many plant sections a mixture of materials will be present. Since very few hazard models can handle mixtures, you will need to select a single representative material. For flammable materials it is generally most appropriate to choose the material whose boiling point is closest to the average normal boiling point of the mixture. For toxic materials you can select the most toxic material, but the initial concentration must be reduced to reflect the concentration in the released material.

Once you have selected the quantities of material, hazard criteria, and representative materials, consequence models can determine the potential hazard zone. Generally, several of the releases will be very similar and it may be possible to reduce the number of modeling runs by grouping similar releases together. The modeling package you choose will provide guidance on how to set up and run the models.

3. Ranking the Facilities

Facilities can be ranked based on the sum of the maximum hazard distances for each release. Only one hazard distance should be used for each release, even if there is the potential for more than one hazard (thermal radiation, explosion overpressure, toxic cloud and flammable vapor cloud). The highest-ranked facility will be the one whose potential releases would reach the greatest total distance.

Keep in mind, however, that ranking facilities on this basis does not account for the impact of the potential hazards on surrounding communities. It may be that your most hazardous facilities are located in remote, unpopulated areas where there is little probability of any injuries outside the plant perimeter.

To factor location into the rankings, multiply the total distances by the average population density in the area surrounding the facility. Where population varies with distance, you may need to vary the density by distance. One effective approach is to consider population density in concentric circles of 1 km, 2 km, 5 km, and 10 km radius.

Table A5.1 Facility Risk Ranking Results

Facility	Hazardous Materials	Maximum Credible Release	Maximum Hazard Distance (miles)	Ranking
Plant A	Anhydrous ammonia	500 lbs/min for 20 minutes	4.7	1
	Propylene	200 lbs/min for 30 minutes; explosion	0.1	
Plant B	Chlorine	2000 lbs in 30 minutes	3.8	2
	Propylene	200 lbs/min for 30 minutes; explosion	0.1	
Plant C	30% aqua ammonia	300 lbs/min for 20 minutes	1.8	3
	Propylene	200 lbs/min for 30 minutes; explosion	0.1	

Note: The maximum hazard distances shown are based on EPA's RMPComp model.

APPENDIX VI: EXAMPLE PRESENTATION ON PSM PLAN

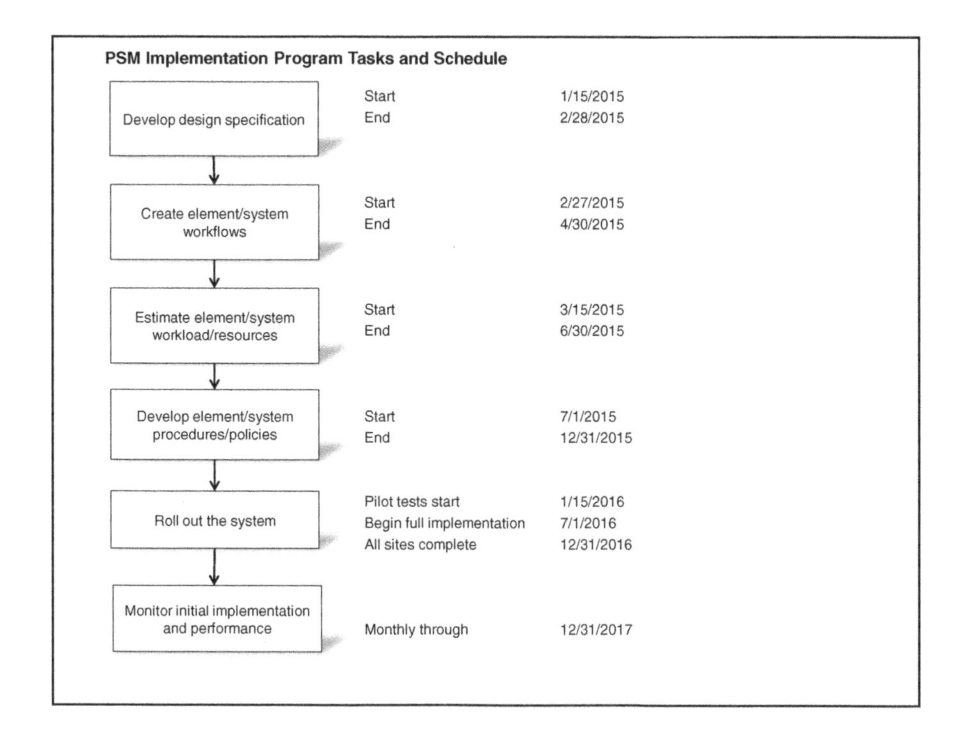

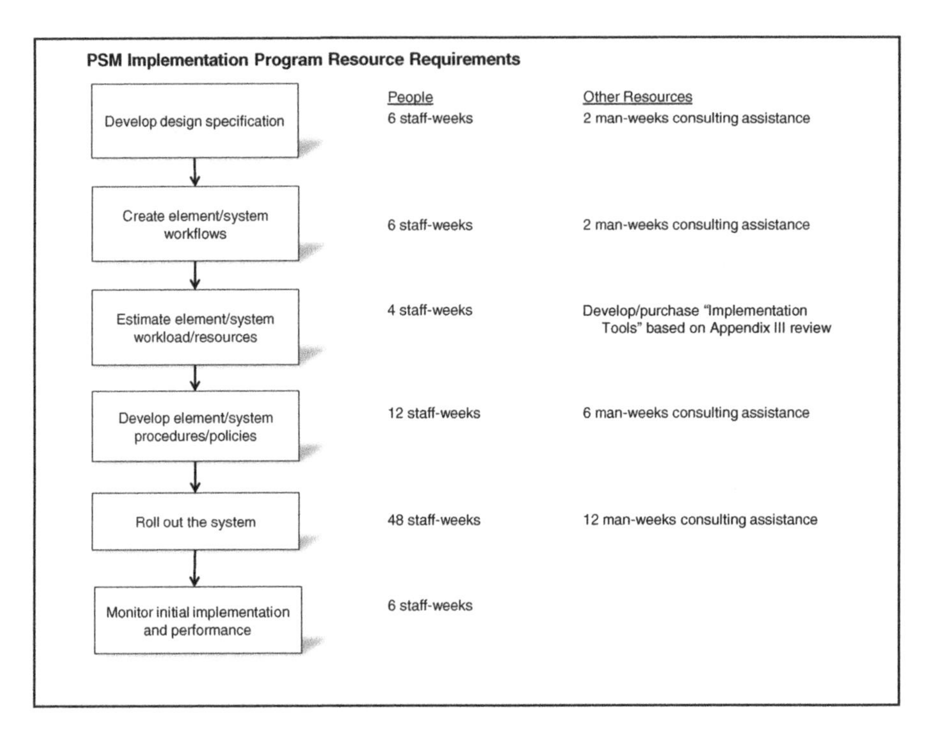
PSM Implementation Program Resource Requirements
People
Other Resources
Develop design specification
6 staff-weeks
2 man-weeks consulting assistance
Create element/system workflows
6 staff-weeks
2 man-weeks consulting assistance
Estimate element/system workload/resources
4 staff-weeks
Develop/purchase "Implementation Tools" based on Appendix III review
Develop element/system procedures/policies
12 staff-weeks
6 man-weeks consulting assistance
Roll out the system
48 staff-weeks
12 man-weeks consulting assistance
Monitor initial implementation and performance
6 staff-weeks

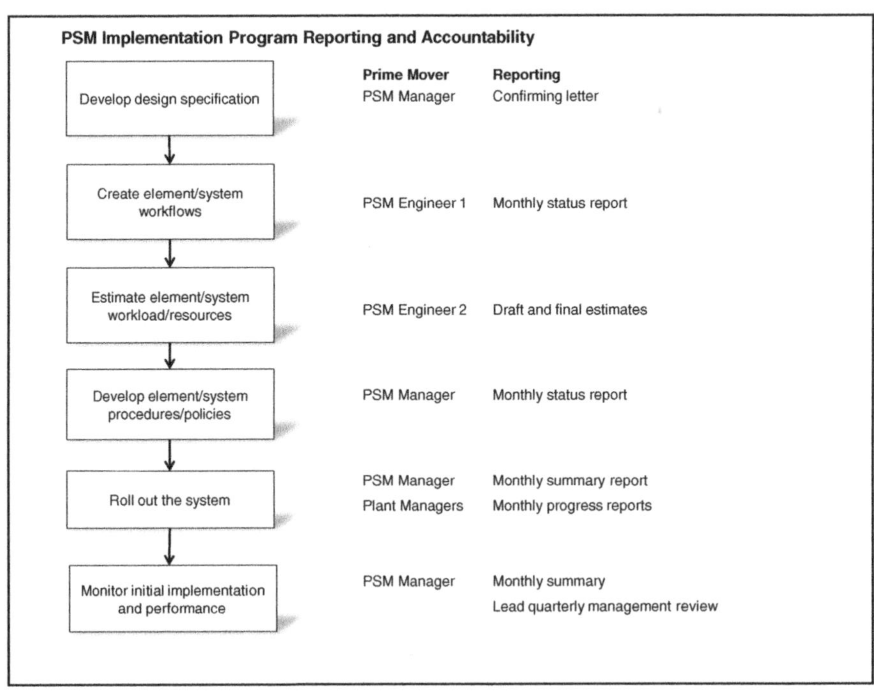
PSM Implementation Program Reporting and Accountability
Prime Mover
Reporting
Develop design specification
PSM Manager
Confirming letter
Create element/system workflows
PSM Engineer 1
Monthly status report
Estimate element/system workload/resources
PSM Engineer 2
Draft and final estimates
Develop element/system procedures/policies
PSM Manager
Monthly status report
Roll out the system
PSM Manager
Monthly summary report
Plant Managers
Monthly progress reports
Monitor initial implementation and performance
PSM Manager
Monthly summary
Lead quarterly management review

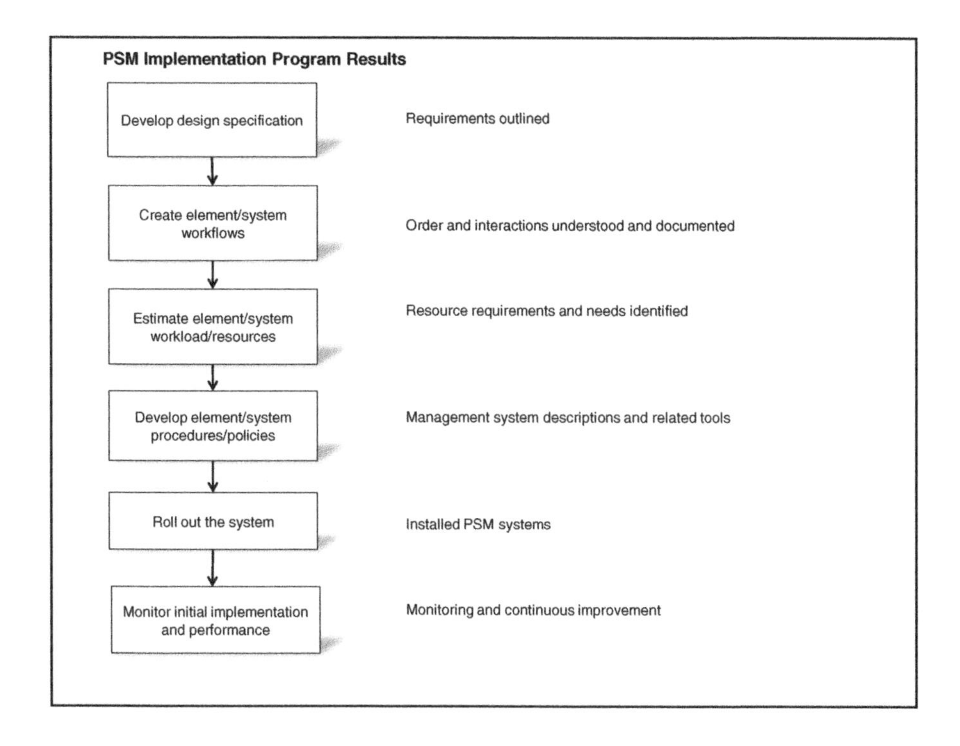
PSM Implementation Program Results
Develop design specification
Requirements outlined
Create element/system workflows
Order and interactions understood and documented
Estimate element/system workload/resources
Resource requirements and needs identified
Develop element/system procedures/policies
Management system descriptions and related tools
Roll out the system
Installed PSM systems
Monitor initial implementation and performance
Monitoring and continuous improvement

APPENDIX VII: MAPPING PERFORMANCE ISSUES TO CULTURE FEATURES

Note: This information is extracted from various ABSG Consulting Inc. presentations, as well as a 2009 poster presentation at the Global Congress on Process Safety.

Sources of data for improving HSE performance

- Incidents
- Performance measures (e.g., metrics, key performance indicators, dashboards)
- Introspective reviews – audits, hazard/risk studies, etc.
- Best practices sharing within industry group
- Benchmarking within peer group
- Global evaluation of state-of-the-art

Terminology

- **Pyramid:** The modified Safety Triangle (or Pyramid). See Chapter 2 for more information.
- **Growth:** Growth in the height (i.e., number) and/or size (i.e., magnitude) of the events at each level of the pyramid
- **Culture – Individual and Organizational Tendencies:** The number/magnitude of poor cultural events (e.g., based on survey scores, interviews, management visits to the processes, housekeeping, evaluation of communication frequency and effectiveness)
- **Unsafe Behaviors and Attitudes:** Number/magnitude of such observations (e.g., based on safety inspections completed, behavior-based safety observations completed)
- **Management System Failures:** Measures of such failures (e.g., based on HSE audit score and findings, number of overdue action items, corrective actions generated, safety meeting attendance, training completed, overdue/delayed MI inspections, alarm frequency, evaluation of incident investigation effectiveness)
- **Precursors:** Indicators of close calls (e.g., based on number of process safety near misses reported, number of unsafe conditions reported, number of demands on safety systems)

- **Incidents:** Measures of “lower level” events (e.g., based on number of first aid incidents, severity rate, small releases and fires, number of Emergency Response Team callouts)
- **Accidents:** Numbers/magnitude of events with significant impact (e.g., based on number of process safety incidents, severity of incidents)

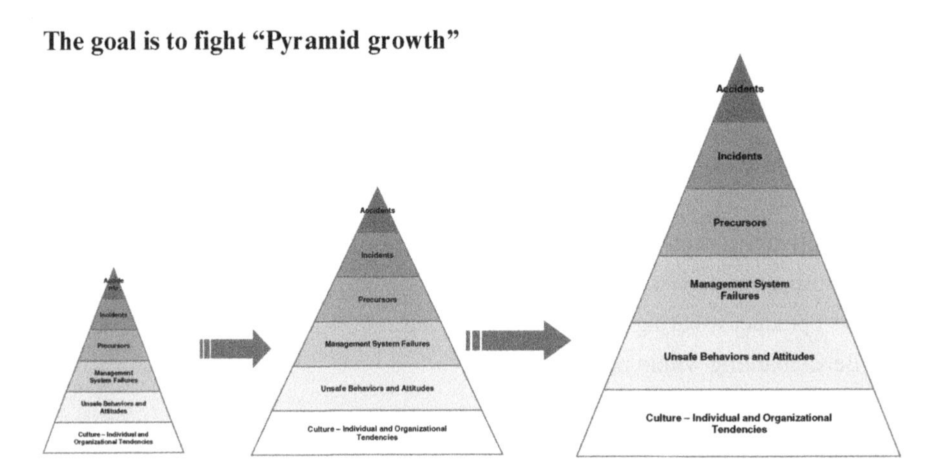

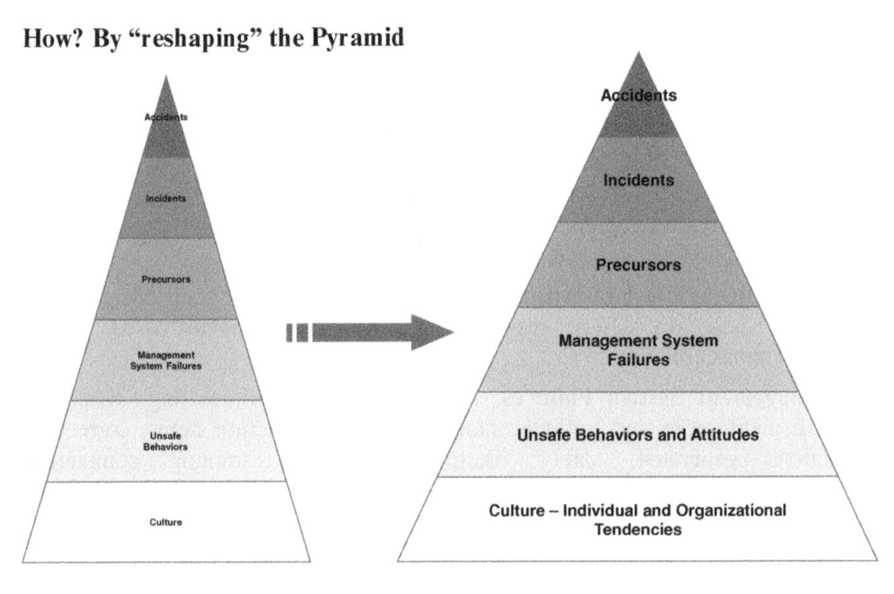

And by "learning lower" on the Pyramid

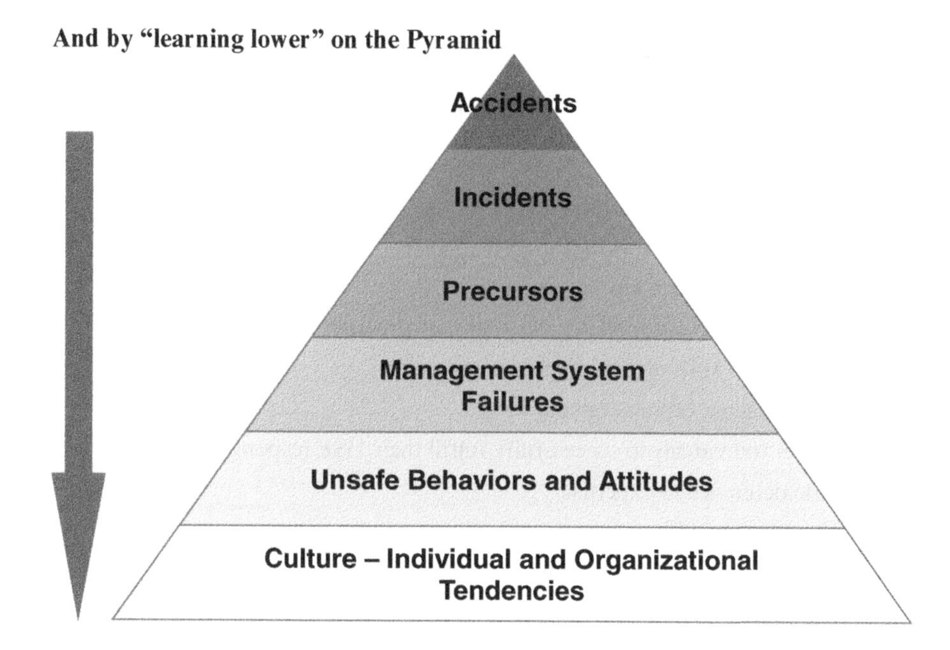

Resulting in a smaller and better shaped Pyramid

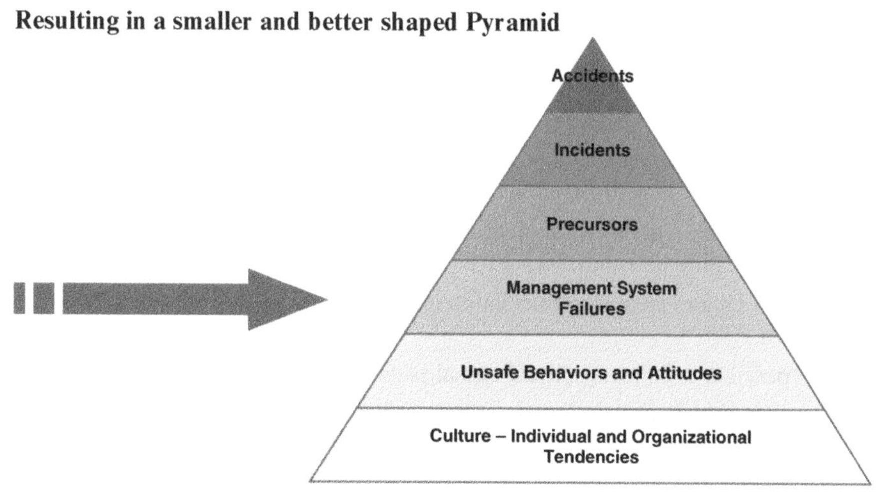

Key points about HSE culture

- Culture is the result of all the actions and inactions in institutional/workforce memory.
- These shape/influence individual behaviors and tendencies.
- Good cultures evolve from common values and attitudes developed as the group seeks to properly solve common problems.

- Culture can be defined as what the people/organization do when no one is looking.
- It is hard to measure and more difficult to change.
- Culture will be the "root cause" of the decade.

Attributes of a good HSE culture

- Espouse HSE as a core value
- Provide strong leadership
- Establish and enforce high standards of performance
- Document the HSE culture emphasis/approach
- Maintain a sense of vulnerability
- Empower individuals to successfully fulfill their HSE responsibilities
- Provide deference to expertise
- Ensure open and effective communications
- Establish a questioning/learning environment (to enhance hazard and risk awareness/understanding)
- Foster mutual trust
- Provide timely response to HSE issues and concerns
- Provide performance monitoring (to promote continuous improvement)

"Connecting the dots" between cultural and technical HSE results/issues

- Audits are one source of information we have to learn from, but they are infrequent and sometimes imprecise.
- Incidents are another source, enabling fixing of root causes and communicating potential risks across the company.
- Cultural factors create the foundation for these HSE results, but are difficult to discern.
- It is possible to connect cultural causal factors and technical HSE outcomes.
- This provides learnings/improvement opportunities nearer the base of the Pyramid.

The process to "connect the dots" looks like this:

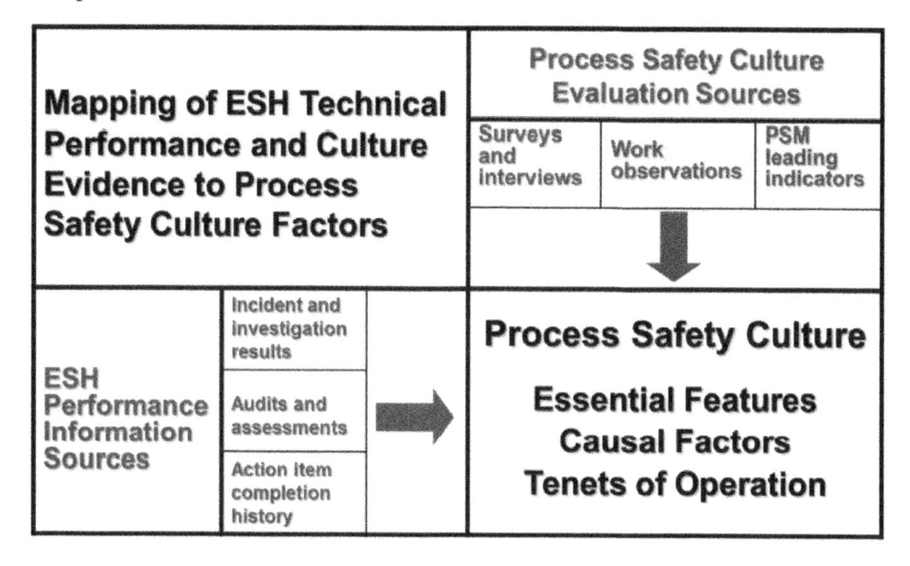

Cultural causal factor analysis approach

- Culture surveys give insight into existing facility/corporate culture, but not how they got there.
- HSE results give insight into possible areas of HSE cultural dysfunction.
- We can relate both results to the essential features of a good HSE culture, or conversely to the cultural causal factors of a "bad" HSE culture.
- Cultural root causes can only be speculated about since forensic analysis of historical culture root causes is difficult – "evidence" is at best anecdotal.
- Tying technical and cultural results together is critical to foster learning and corrective action.

Thus, evaluating the process safety/HSE culture at a facility/company should include:

- Culture survey/interviews
    - Determine focus – safety, process safety, or HSE
    - Develop survey
    - Pilot test
    - Decide on delivery/administration method – paper, online
    - Collect/analyze data

- Management interviews
    - Determine focus – safety, process safety, or HSE
    - Develop interview question set
    - Perform interviews
    - Collect/analyze data
- HSE management system performance measures
    - Identify existing metrics or define data to collect
    - Analyze results

The Process Safety Performance Assurance Review (PAR)© Strategy picture on the next page summarizes all of the above information.

Connecting the Dots

Process Safety Performance Assurance Review (PAR)© Strategy

An organization's CULTURE can influence its performance, productivity, profitability and risk. Process safety is more than a program - it's an integral part of the culture that DRIVES your business. How do you find out how your organization's culture is affecting your PROCESS SAFETY risk profile?

Process Safety/ESH Culture Evaluation Sources

- Surveys and interviews
- Work observations
- PSM/EHS leading indicators

Process Safety/ESH Performance Information Sources

- Incidents and investigation results
- Audits and assessments
- Action item completion history

ABS Consulting

AN ABS GROUP COMPANY

INDEX

Printed and bound by CPI Group (UK) Ltd, Croydon, CR0 4YY

23/04/2025

14660910-0002